华章IT | Information Technology

智能系统与技术丛书

TensorFlow Machine Learning Cookbook
Second Edition

TensorFlow机器学习
实战指南

（原书第2版）

［美］尼克·麦克卢尔（Nick McClure）著

李飞 刘凯 卢建华 李静 赵秀丽 译

机械工业出版社
China Machine Press

图书在版编目（CIP）数据

TensorFlow 机器学习实战指南（原书第 2 版）/（美）尼克·麦克卢尔（Nick McClure）著；李飞等译 . —北京：机械工业出版社，2019.7

（智能系统与技术丛书）

书名原文：TensorFlow Machine Learning Cookbook, Second Edition

ISBN 978-7-111-63126-2

I. T… II. ①尼… ②李… III. ①人工智能 – 算法 – 研究 ②机器学习 IV. TP18

中国版本图书馆 CIP 数据核字（2019）第 138401 号

本书版权登记号：图字 01-2018-8343

Nick McClure：TensorFlow Machine Learning Cookbook, Second Edition (ISBN: 978-1-78913-168-0).

Copyright © 2018 Packt Publishing. First published in the English language under the title "TensorFlow Machine Learning Cookbook, Second Edition".

All rights reserved.

Chinese simplified language edition published by China Machine Press.

Copyright © 2019 by China Machine Press.

本书中文简体字版由 Packt Publishing 授权机械工业出版社独家出版。未经出版者书面许可，不得以任何方式复制或抄袭本书内容。

TensorFlow 机器学习实战指南（原书第 2 版）

出版发行：机械工业出版社（北京市西城区百万庄大街 22 号 邮政编码：100037）			
责任编辑：刘 锋		责任校对：殷 虹	
印　　刷：大厂回族自治县益利印刷有限公司		版　　次：2019 年 7 月第 1 版第 1 次印刷	
开　　本：186mm×240mm　1/16		印　　张：18.5	
书　　号：ISBN 978-7-111-63126-2		定　　价：89.00 元	

凡购本书，如有缺页、倒页、脱页，由本社发行部调换

客服热线：(010) 88379426 88361066　　　　　投稿热线：(010) 88379604

购书热线：(010) 68326294　　　　　　　　　　读者信箱：hzit@hzbook.com

版权所有 • 侵权必究

封底无防伪标均为盗版

本书法律顾问：北京大成律师事务所　韩光／邹晓东

THE TRANSLATOR'S WORDS
译 者 序

经过了几年的发展，TensorFlow 已经成为最受欢迎的开源机器学习框架之一，活跃的社区也加速了它的普及程度。它快速、灵活并能够大规模部署于工业生产环境，让每个开发者和研究者都能够方便地使用人工智能来解决多种多样的问题；并且它由谷歌开发、维护，有力地保障了其持续性支持与开发。

原著第 1 版颇受欢迎，第 2 版在第 1 版的基础之上做了一些微调，并在第 10 章与第 11 章增加了部分内容，进一步完善了本书的体系。第 1 版翻译水平颇高，为第 2 版的翻译打下了良好的基础。

本书第 1 章和第 2 章介绍了关于 TensorFlow 使用的基础知识，后续章节则针对一些典型算法和典型应用场景进行了实现，并配有较详细的程序说明，可读性非常强。读者如果能对其中代码进行复现，则必定会对 TensorFlow 的使用了如指掌。

本书翻译工作由海军航空大学和鲁东大学的机器学习研究者共同完成。其中第 1～5 章由刘凯博士完成，第 6 章由赵秀丽副教授完成，第 7 章由卢建华教授完成，第 8～11 章由李飞博士完成，鲁东大学的李静老师对全书进行了翻译风格上的统一和纠错。特别感谢刘锦涛博士对本书翻译工作的关心与支持，感谢第 1 版译者曾益强老师打下的基础。

由于译者水平有限，错误和不妥之处在所难免，衷心希望各位读者批评指正。

2019 年 3 月

译者

ABOUT THE REVIEWER
审校者简介

　　Sujit Pal 是 Elsevier 实验室的技术研究专家，致力于语义搜索、自然语言处理、机器学习和深度学习。在 Elsevier，他参与了多项创新性研究工作，包括搜索质量度量与改进、图像分类、重复检测，以及医学和科技语料库的标注和本体开发。他与 Antonio Gulli 合著过深度学习的著作并长期在他的博客 Salmon Run 上分享技术；同时他也是《*Reinforcement Learning in Python*》的审校者。

PREFACE

前　　言

2015 年 11 月，Google 公司开源 TensorFlow，随后不久 TensorFlow 成为 GitHub 上最受欢迎的机器学习库。TensorFlow 创建计算图、自动求导和定制化的方式使得其能够很好地解决许多不同的机器学习算法问题。

本书介绍了许多机器学习算法，将其应用到真实场景和数据中，并解释产生的结果。

本书的主要内容

第 1 章介绍 TensorFlow 的基本概念，包括张量、变量和占位符；同时展示了在 TensorFlow 中如何使用矩阵和各种数学运算。本章末尾讲述如何访问本书所需的数据源。

第 2 章介绍如何在计算图中连接第 1 章介绍的所有算法组件，创建一个简单的分类器。接着，介绍计算图、损失函数、反向传播和训练模型。

第 3 章重点讨论使用 TensorFlow 实现各种线性回归算法，比如，戴明回归、lasso 回归、岭回归、弹性网络回归和逻辑回归，也展示了如何在 TensorFlow 计算图中实现每种回归算法。

第 4 章介绍支持向量机（SVM）算法，展示如何在 TensorFlow 中实现线性 SVM 算法、非线性 SVM 算法和多分类 SVM 算法。

第 5 章展示如何使用数值度量、文本度量和归一化距离函数实现最近邻域法。我们使用最近邻域法进行地址间的记录匹配和 MNIST 数据库中手写数字的分类。

第 6 章讲述如何使用 TensorFlow 实现神经网络算法，包括操作门和激励函数的概念。随后展示一个简单的神经网络并讨论如何建立不同类型的神经网络层。本章末尾通过神经网络算法教 TensorFlow 玩井字棋游戏。

第 7 章阐述借助 TensorFlow 实现的各种文本处理算法。我们展示如何实现文本的"词袋"和 TF-IDF 算法。然后介绍 CBOW 和 skip-gram 模型的神经网络文本表示方式，并对于 Word2Vec 和 Doc2Vec 用这些方法来做预测，例如预测一个文本消息是否为垃圾信息。

第 8 章扩展神经网络算法，说明如何借助卷积神经网络（CNN）算法在图像上应用神经网络算法。我们展示如何构建一个简单的 CNN 进行 MNIST 数字识别，并扩展到 CIFAR-10 任务中的彩色图片，也阐述了如何针对自定义任务扩展之前训练的图像识别模型。本章末尾详细解释 TensorFlow 实现的图像风格和 Deep-Dream 算法。

第 9 章解释在 TensorFlow 中如何实现循环神经网络（RNN）算法，展示如何进行垃圾邮件预测和在莎士比亚文本样本集上扩展 RNN 模型生成文本。接着训练 Seq2Seq 模型实现德语 – 英语的翻译。本章末尾展示如何用孪生 RNN 模型进行地址记录匹配。

第 10 章介绍 TensorFlow 产品级用例和开发提示，同时介绍如何利用多处理设备（比如，GPU）和在多个设备上实现分布式 TensorFlow。

第 11 章展示 TensorFlow 如何实现 k-means 算法、遗传算法和求解常微分方程（ODE），还介绍了 Tensorboard 的各种用法和如何查看计算图指标。

阅读本书前的准备

书中的章节都会使用 TensorFlow，其官网为 https://www.tensorflow.org/，它是基于 Python 3（https://www.python.org/downloads/）编写的。大部分章节需要访问从网络中下载的数据集。

本书的目标读者

本书主要是为了帮助那些想要同时了解 TensorFlow 和主流机器学习算法应用策略的业余爱好者、程序员和机器学习发烧友。阅读本书，首先要有基本的数学知识和 Python 编程技巧。本书的主要目的在于介绍 TensorFlow 和提供基于 TensorFlow 的各种机器学习算法优秀案例，不涉及数学、机器学习以及 Python 编程的具体问题。更为恰当的描述是，本书是对这三个方面的宏观介绍。鉴于这一原因，读者可能会觉得本书中有的内容过于简单，有的过于烦琐。如果读者具有坚实的机器学习基础，通常会觉得书中的 TensorFlow 代码对自己帮助较大；如果读者善于编写 Python 程序，那么可能会对书中的代码注释感兴趣。如果读者想要深入研究某些特定领域，可以从许多章节末尾的"延伸学习"部分所提供的参考文献和资源中了解更多信息。

模块说明

在本书中，你会频繁看到开始、动手做、工作原理、延伸学习和参考这几个模块。

为了系统地学习相关技术，下面简单解释一下：

❑ 开始

该节告诉读者该技术的内容，描述如何准备软件或者前期的准备工作。

❑ 动手做

具体的操作步骤。

❑ 工作原理

详细解释前一节发生了什么。

❑ 延伸学习

附加资源，以供读者延伸学习。

❑ 参考

提供有用的链接和有帮助的资源信息。

下载示例代码

读者可登录华章网站（www.hzbook.com）下载本书示例代码文件。

目　录

译者序
审校者简介
前言

第 1 章　TensorFlow 基础 ……… 1
1.1　简介 ……………………………… 1
1.2　TensorFlow 如何工作 …………… 1
　　1.2.1　开始 ……………………… 1
　　1.2.2　动手做 …………………… 2
　　1.2.3　工作原理 ………………… 3
　　1.2.4　参考 ……………………… 3
1.3　声明变量和张量 ………………… 4
　　1.3.1　开始 ……………………… 4
　　1.3.2　动手做 …………………… 4
　　1.3.3　工作原理 ………………… 6
　　1.3.4　延伸学习 ………………… 6
1.4　使用占位符和变量 ……………… 6
　　1.4.1　开始 ……………………… 6
　　1.4.2　动手做 …………………… 6
　　1.4.3　工作原理 ………………… 7
　　1.4.4　延伸学习 ………………… 7
1.5　操作（计算）矩阵 ……………… 8
　　1.5.1　开始 ……………………… 8
　　1.5.2　动手做 …………………… 8
　　1.5.3　工作原理 ………………… 10
1.6　声明操作 ………………………… 10
　　1.6.1　开始 ……………………… 10
　　1.6.2　动手做 …………………… 10
　　1.6.3　工作原理 ………………… 12
　　1.6.4　延伸学习 ………………… 12
1.7　实现激励函数 …………………… 12
　　1.7.1　开始 ……………………… 12
　　1.7.2　动手做 …………………… 12
　　1.7.3　工作原理 ………………… 14
　　1.7.4　延伸学习 ………………… 14
1.8　读取数据源 ……………………… 14
　　1.8.1　开始 ……………………… 15
　　1.8.2　动手做 …………………… 15
　　1.8.3　工作原理 ………………… 18
　　1.8.4　参考 ……………………… 18
1.9　其他资源 ………………………… 19
　　1.9.1　开始 ……………………… 19
　　1.9.2　动手做 …………………… 19

第 2 章　TensorFlow 进阶 ……… 20
2.1　简介 ……………………………… 20

- 2.2 计算图中的操作 ·············· 20
  - 2.2.1 开始 ··················· 20
  - 2.2.2 动手做 ··············· 21
  - 2.2.3 工作原理 ··········· 21
- 2.3 TensorFlow 的嵌入 Layer ····· 21
  - 2.3.1 开始 ··················· 21
  - 2.3.2 动手做 ··············· 22
  - 2.3.3 工作原理 ··········· 22
  - 2.3.4 延伸学习 ··········· 22
- 2.4 TensorFlow 的多层 Layer ····· 23
  - 2.4.1 开始 ··················· 23
  - 2.4.2 动手做 ··············· 23
  - 2.4.3 工作原理 ··········· 24
- 2.5 TensorFlow 实现损失函数 ··· 24
  - 2.5.1 开始 ··················· 25
  - 2.5.2 动手做 ··············· 26
  - 2.5.3 工作原理 ··········· 28
  - 2.5.4 延伸学习 ··········· 28
- 2.6 TensorFlow 实现反向传播 ··· 29
  - 2.6.1 开始 ··················· 29
  - 2.6.2 动手做 ··············· 30
  - 2.6.3 工作原理 ··········· 33
  - 2.6.4 延伸学习 ··········· 33
  - 2.6.5 参考 ··················· 33
- 2.7 TensorFlow 实现批量训练和随机训练 ····················· 34
  - 2.7.1 开始 ··················· 34
  - 2.7.2 动手做 ··············· 34
  - 2.7.3 工作原理 ··········· 35
  - 2.7.4 延伸学习 ··········· 36
- 2.8 TensorFlow 实现创建分类器 ··· 36
  - 2.8.1 开始 ··················· 36
  - 2.8.2 动手做 ··············· 37
  - 2.8.3 工作原理 ··········· 38
  - 2.8.4 延伸学习 ··········· 39
  - 2.8.5 参考 ··················· 39
- 2.9 TensorFlow 实现模型评估 ··· 39
  - 2.9.1 开始 ··················· 39
  - 2.9.2 动手做 ··············· 40
  - 2.9.3 工作原理 ··········· 43

第 3 章 基于 TensorFlow 的线性回归 ·············· 44

- 3.1 简介 ·························· 44
- 3.2 用 TensorFlow 求逆矩阵 ····· 44
  - 3.2.1 开始 ··················· 45
  - 3.2.2 动手做 ··············· 45
  - 3.2.3 工作原理 ··········· 46
- 3.3 用 TensorFlow 实现矩阵分解 ··· 46
  - 3.3.1 开始 ··················· 46
  - 3.3.2 动手做 ··············· 46
  - 3.3.3 工作原理 ··········· 47
- 3.4 用 TensorFlow 实现线性回归算法 ··························· 47
  - 3.4.1 开始 ··················· 48
  - 3.4.2 动手做 ··············· 48
  - 3.4.3 工作原理 ··········· 50
- 3.5 理解线性回归中的损失函数 ··· 51
  - 3.5.1 开始 ··················· 51
  - 3.5.2 动手做 ··············· 51
  - 3.5.3 工作原理 ··········· 52
  - 3.5.4 延伸学习 ··········· 53

3.6 用 TensorFlow 实现戴明回归
算法 ································· 53
　3.6.1　开始 ························· 54
　3.6.2　动手做 ······················ 54
　3.6.3　工作原理 ··················· 55
3.7 用 TensorFlow 实现 lasso 回归
和岭回归算法 ···················· 56
　3.7.1　开始 ························· 56
　3.7.2　动手做 ······················ 56
　3.7.3　工作原理 ··················· 58
　3.7.4　延伸学习 ··················· 58
3.8 用 TensorFlow 实现弹性网络
回归算法 ··························· 58
　3.8.1　开始 ························· 58
　3.8.2　动手做 ······················ 58
　3.8.3　工作原理 ··················· 60
3.9 用 TensorFlow 实现逻辑回归
算法 ································· 60
　3.9.1　开始 ························· 60
　3.9.2　动手做 ······················ 61
　3.9.3　工作原理 ··················· 63

第 4 章　基于 TensorFlow 的支持向量机

4.1 简介 ································· 65
4.2 线性支持向量机的使用 ······ 67
　4.2.1　开始 ························· 67
　4.2.2　动手做 ······················ 67
　4.2.3　工作原理 ··················· 70
4.3 弱化为线性回归 ··············· 71
　4.3.1　开始 ························· 71
　4.3.2　动手做 ······················ 72

4.3.3　工作原理 ··················· 74
4.4 TensorFlow 上核函数的使用 ······ 75
　4.4.1　开始 ························· 75
　4.4.2　动手做 ······················ 76
　4.4.3　工作原理 ··················· 80
　4.4.4　延伸学习 ··················· 80
4.5 用 TensorFlow 实现非线性支持
向量机 ······························· 80
　4.5.1　开始 ························· 80
　4.5.2　动手做 ······················ 80
　4.5.3　工作原理 ··················· 83
4.6 用 TensorFlow 实现多类支持
向量机 ······························· 83
　4.6.1　开始 ························· 83
　4.6.2　动手做 ······················ 84
　4.6.3　工作原理 ··················· 87

第 5 章　最近邻域法

5.1 简介 ································· 88
5.2 最近邻域法的使用 ············ 89
　5.2.1　开始 ························· 89
　5.2.2　动手做 ······················ 89
　5.2.3　工作原理 ··················· 92
　5.2.4　延伸学习 ··················· 92
5.3 如何度量文本距离 ············ 92
　5.3.1　开始 ························· 93
　5.3.2　动手做 ······················ 93
　5.3.3　工作原理 ··················· 95
　5.3.4　延伸学习 ··················· 95
5.4 用 TensorFlow 实现混合距离
计算 ································· 95
　5.4.1　开始 ························· 96

5.4.2　动手做 ································· 96
　　5.4.3　工作原理 ····························· 98
　　5.4.4　延伸学习 ····························· 98
5.5　用 TensorFlow 实现地址
　　匹配 ··· 99
　　5.5.1　开始 ···································· 99
　　5.5.2　动手做 ································· 99
　　5.5.3　工作原理 ··························· 101
5.6　用 TensorFlow 实现图像
　　识别 ··· 102
　　5.6.1　开始 ·································· 102
　　5.6.2　动手做 ······························· 102
　　5.6.3　工作原理 ··························· 104
　　5.6.4　延伸学习 ··························· 105

第 6 章　神经网络算法 ··············· 106
6.1　简介 ··· 106
6.2　用 TensorFlow 实现门函数 ········ 107
　　6.2.1　开始 ·································· 107
　　6.2.2　动手做 ······························· 108
　　6.2.3　工作原理 ··························· 110
6.3　使用门函数和激励函数 ············ 110
　　6.3.1　开始 ·································· 111
　　6.3.2　动手做 ······························· 111
　　6.3.3　工作原理 ··························· 113
　　6.3.4　延伸学习 ··························· 113
6.4　用 TensorFlow 实现单层神经
　　网络 ··· 114
　　6.4.1　开始 ·································· 114
　　6.4.2　动手做 ······························· 114
　　6.4.3　工作原理 ··························· 116
　　6.4.4　延伸学习 ··························· 117

6.5　用 TensorFlow 实现神经网络
　　常见层 ··· 117
　　6.5.1　开始 ·································· 117
　　6.5.2　动手做 ······························· 117
　　6.5.3　工作原理 ··························· 122
6.6　用 TensorFlow 实现多层神经
　　网络 ··· 123
　　6.6.1　开始 ·································· 123
　　6.6.2　动手做 ······························· 123
　　6.6.3　工作原理 ··························· 127
6.7　线性预测模型的优化 ················ 128
　　6.7.1　开始 ·································· 128
　　6.7.2　动手做 ······························· 128
　　6.7.3　工作原理 ··························· 131
6.8　用 TensorFlow 基于神经网络
　　实现井字棋 ··································· 132
　　6.8.1　开始 ·································· 133
　　6.8.2　动手做 ······························· 134
　　6.8.3　工作原理 ··························· 139

第 7 章　自然语言处理 ··············· 140
7.1　简介 ··· 140
7.2　词袋的使用 ································· 141
　　7.2.1　开始 ·································· 141
　　7.2.2　动手做 ······························· 142
　　7.2.3　工作原理 ··························· 146
　　7.2.4　延伸学习 ··························· 146
7.3　用 TensorFlow 实现 TF-IDF
　　算法 ··· 146
　　7.3.1　开始 ·································· 146
　　7.3.2　动手做 ······························· 147
　　7.3.3　工作原理 ··························· 150

7.3.4 延伸学习 ……………… 151	8.2.4 延伸学习 ……………… 182
7.4 用 TensorFlow 实现 skip-gram	8.2.5 参考 …………………… 183
模型 ……………………………… 151	8.3 用 TensorFlow 实现进阶的
7.4.1 开始 ………………… 151	CNN …………………………… 183
7.4.2 动手做 ……………… 152	8.3.1 开始 ………………… 183
7.4.3 工作原理 …………… 158	8.3.2 动手做 ……………… 183
7.4.4 延伸学习 …………… 158	8.3.3 工作原理 …………… 189
7.5 用 TensorFlow 实现 CBOW 词	8.3.4 参考 ………………… 190
嵌入模型 ……………………… 158	8.4 再训练已有的 CNN 模型 …… 190
7.5.1 开始 ………………… 158	8.4.1 开始 ………………… 190
7.5.2 动手做 ……………… 159	8.4.2 动手做 ……………… 191
7.5.3 工作原理 …………… 163	8.4.3 工作原理 …………… 193
7.5.4 延伸学习 …………… 163	8.4.4 参考 ………………… 193
7.6 使用 TensorFlow 的 Word2Vec	8.5 用 TensorFlow 实现图像风格
预测 …………………………… 163	迁移 …………………………… 193
7.6.1 开始 ………………… 163	8.5.1 开始 ………………… 194
7.6.2 动手做 ……………… 163	8.5.2 动手做 ……………… 194
7.6.3 工作原理 …………… 168	8.5.3 工作原理 …………… 199
7.6.4 延伸学习 …………… 168	8.5.4 参考 ………………… 199
7.7 用 TensorFlow 实现基于 Doc2Vec	8.6 用 TensorFlow 实
的情感分析 …………………… 168	现 DeepDream ………………… 199
7.7.1 开始 ………………… 168	8.6.1 开始 ………………… 199
7.7.2 动手做 ……………… 169	8.6.2 动手做 ……………… 199
7.7.3 工作原理 …………… 175	8.6.3 延伸学习 …………… 204
第 8 章 卷积神经网络 …………… 176	8.6.4 参考 ………………… 204
8.1 简介 ………………………… 176	**第 9 章 循环神经网络** …………… 205
8.2 用 TensorFlow 实现简单的	9.1 简介 ………………………… 205
CNN …………………………… 177	9.2 用 TensorFlow 实现 RNN 模型
8.2.1 开始 ………………… 177	进行垃圾邮件预测 …………… 206
8.2.2 动手做 ……………… 177	9.2.1 开始 ………………… 206
8.2.3 工作原理 …………… 182	9.2.2 动手做 ……………… 206

XIII

9.2.3 工作原理 ………………… 211
9.2.4 延伸学习 ………………… 211
9.3 用 TensorFlow 实现 LSTM
模型 ……………………………… 211
 9.3.1 开始 ……………………… 211
 9.3.2 动手做 …………………… 212
 9.3.3 工作原理 ………………… 218
 9.3.4 延伸学习 ………………… 218
9.4 TensorFlow 堆叠多层 LSTM … 219
 9.4.1 开始 ……………………… 219
 9.4.2 动手做 …………………… 219
 9.4.3 工作原理 ………………… 221
9.5 用 TensorFlow 实现 Seq2Seq
翻译模型 ………………………… 221
 9.5.1 开始 ……………………… 221
 9.5.2 动手做 …………………… 222
 9.5.3 工作原理 ………………… 232
 9.5.4 延伸学习 ………………… 232
9.6 TensorFlow 训练孪生 RNN
度量相似度 ……………………… 232
 9.6.1 开始 ……………………… 232
 9.6.2 动手做 …………………… 233
 9.6.3 延伸学习 ………………… 238

第 10 章 TensorFlow 产品化 …… 239

10.1 简介 ……………………………… 239
10.2 TensorFlow 的单元测试 ……… 239
 10.2.1 开始 …………………… 239
 10.2.2 工作原理 ……………… 244
10.3 TensorFlow 的多设备使用 …… 244
 10.3.1 开始 …………………… 244
 10.3.2 动手做 ………………… 245

10.3.3 工作原理 ……………… 246
10.3.4 延伸学习 ……………… 246
10.4 分布式 TensorFlow 实践 ……… 246
 10.4.1 开始 …………………… 247
 10.4.2 动手做 ………………… 247
 10.4.3 工作原理 ……………… 248
10.5 TensorFlow 产品化开发提示 … 248
 10.5.1 开始 …………………… 248
 10.5.2 动手做 ………………… 248
 10.5.3 工作原理 ……………… 250
10.6 TensorFlow 产品化的实例 …… 250
 10.6.1 开始 …………………… 250
 10.6.2 动手做 ………………… 250
 10.6.3 工作原理 ……………… 253
10.7 TensorFlow 服务部署 ………… 253
 10.7.1 开始 …………………… 253
 10.7.2 动手做 ………………… 253
 10.7.3 工作原理 ……………… 256
 10.7.4 延伸学习 ……………… 257

第 11 章 TensorFlow 的进阶
应用 ……………………… 258

11.1 简介 ……………………………… 258
11.2 TensorFlow 可视化：
Tensorboard ……………………… 258
 11.2.1 开始 …………………… 258
 11.2.2 动手做 ………………… 259
 11.2.3 延伸学习 ……………… 261
11.3 用 TensorFlow 实现遗传算法 … 263
 11.3.1 开始 …………………… 263
 11.3.2 动手做 ………………… 264
 11.3.3 工作原理 ……………… 266

11.3.4 延伸学习 …… 266
11.4 用 TensorFlow 实现 k-means 聚类算法 …… 267
    11.4.1 开始 …… 267
    11.4.2 动手做 …… 267
    11.4.3 延伸学习 …… 270
11.5 用 TensorFlow 求解常微分方程组 …… 270
    11.5.1 开始 …… 271
    11.5.2 动手做 …… 271
    11.5.3 工作原理 …… 272
    11.5.4 参考 …… 272
11.6 用 TensorFlow 实现随机森林算法 …… 273
    11.6.1 开始 …… 273
    11.6.2 动手做 …… 273
    11.6.3 工作原理 …… 276
    11.6.4 参考 …… 276
11.7 将 Keras 作为 TensorFlow API 使用 …… 277
    11.7.1 开始 …… 277
    11.7.2 动手做 …… 277
    11.7.3 工作原理 …… 280
    11.7.4 参考 …… 281

CHAPTER 1

第 1 章

TensorFlow 基础

本章将介绍 TensorFlow 的基本概念，帮助读者去理解 TensorFlow 是如何工作的，以及它如何访问数据集和其他资源。学完本章可以掌握以下知识点：
- TensorFlow 如何工作
- 声明变量和张量
- 使用占位符和变量
- 操作（计算）矩阵
- 声明操作
- 实现激励函数
- 读取数据源
- 其他资源

1.1 简介

Google 的 TensorFlow 引擎提供了一种解决机器学习问题的高效方法。机器学习在各行各业应用广泛，特别是计算机视觉、语音识别、语言翻译和健康医疗等领域。本书将详细介绍 TensorFlow 操作的基本步骤以及代码。这些基础知识对理解本书后续章节非常有用。

1.2 TensorFlow 如何工作

首先，TensorFlow 的计算看起来并不是很复杂，因为 TensorFlow 的计算过程和算法开发相当容易。这章将引导读者理解 TensorFlow 算法的伪代码。

1.2.1 开始

截至目前，TensorFlow 支持 Linux、macOS 和 Windows 操作系统。本书的代码都是在 Linux 操作系统上实现和运行的，不过运行在其他操作系统上也没问题。本书的代码可以在

GitHub（https://github.com/nfmcclure/tensorflow_cookbook）或 Packt 代码库（https://github.com/PacktPublishing/TensorFlow-Machine-Learning-Cookbook-Second-Edition）获取。虽然 TensorFlow 是用 C++ 编写，但是全书只介绍 TensorFlow 的 Python 使用方式。本书将使用 Python 3.6+（https://www.python.org）和 TensorFlow 1.10.0+（https://www.tensorflow.org）。尽管 TensorFlow 能在 CPU 上运行，但大部分算法在 GPU 上会运行得更快，它支持英伟达显卡（Nvidia Compute Capability v4.0+，推荐 v5.1）。TensorFlow 上常用的 GPU 是英伟达特斯拉（Nvidia Tesla）和英伟达帕斯卡（Nvidia Pascal），至少需要 4GB 的视频 RAM。为了在 GPU 上运行，需要下载 Nvidia Cuda Toolkit 及其 v5.x+（https://developer.nvidia.com/cuda-downloads）。本书还依赖 Python 的包：Scipy、Numpy 和 Scikit-Learn，这些包均包含在 Anaconda 中（https://www.continuum.io/downloads）。

1.2.2 动手做

这里是 TensorFlow 算法的一般流程，本书提炼出的纲领如下：

1. 导入 / 生成样本数据集：所有的机器学习算法都依赖样本数据集，本书的数据集既有生成的样本数据集，也有外部公开的样本数据集。有时，生成的数据集会更容易符合预期结果，但是本书大部分都是访问外部公开的样本数据集，具体细节见 1.8 节。

2. 转换和归一化数据：一般来讲，输入样本数据集并不符合 TensorFlow 期望的形状，所以需要转换数据格式以满足 TensorFlow。当数据集的维度或者类型不符合所用机器学习算法的要求时，需要在使用前进行数据转换。大部分机器学习算法期待的输入样本数据是归一化的数据。TensorFlow 具有内建函数来归一化数据，如下：

```
import tensorflow as tf
data = tf.nn.batch_norm_with_global_normalization(...)
```

3. 划分样本数据集为训练样本集、测试样本集和验证样本集：一般要求机器学习算法的训练样本集和测试样本集是不同的数据集。另外，许多机器学习算法要求超参数调优，所以需要验证样本集来决定最优的超参数。

4. 设置机器学习参数（超参数）：机器学习经常要有一系列的常量参数。例如，迭代次数、学习率或者其他固定参数。约定俗成的习惯是一次性初始化所有的机器学习参数，从而提高程序的可读性，读者经常看到的形式如下：

```
learning_rate = 0.01
batch_size = 100
iterations = 1000
```

5. 初始化变量和占位符：在求解最优化过程中（最小化损失函数），TensorFlow 通过占位符传入数据，并调整变量（权重 / 偏差）。TensorFlow 指定数据大小和数据类型来初始化变量和占位符。本书大部分使用 float32 数据类型，TensorFlow 也支持 float64 和 float16。注意，使用的数据类型字节数越多结果越精确，同时运行速度越慢。使用方式如下：

```
a_var = tf.constant(42)
x_input = tf.placeholder(tf.float32, [None, input_size])
y_input = tf.placeholder(tf.float32, [None, num_classes])
```

6. 定义模型结构：在获取样本数据集、初始化变量和占位符后，开始定义机器学习模型。TensorFlow通过选择操作、变量和占位符的值来构建计算图，详细讲解见2.2节。这里给出简单的线性模型 $y = mx + b$：

```
y_pred = tf.add(tf.mul(x_input, weight_matrix), b_matrix)
```

7. 声明损失函数：定义完模型后，需要声明损失函数来评估输出结果。损失函数能说明预测值与实际值的差距，损失函数的种类将在第2章详细展示，这里给出 n 个样本的均方误差（$\text{loss} = (1/n)\Sigma(y_{实际} - y_{预测})^2$）：

```
loss = tf.reduce_mean(tf.square(y_actual - y_pred))
```

8. 初始化模型和训练模型：TensorFlow创建计算图实例，通过占位符传入数据，维护变量的状态信息。下面是初始化计算图的一种方式：

```
with tf.Session(graph=graph) as session:
...
session.run(...)
...
```

也可以用如下的方式初始化计算图：

```
session = tf.Session(graph=graph)
session.run(...)
```

9. 评估机器学习模型：一旦构建计算图，并训练机器学习模型后，需要寻找某种标准来评估机器学习模型对新样本数据集的效果。通过对训练样本集和测试样本集的评估，可以确定机器学习模型是过拟合还是欠拟合。这些将在后续章节来解决。

10. 调优超参数：大部分情况下，机器学习者需要基于模型效果来回调整一些超参数。通过调整不同的超参数来重复训练模型，并用验证样本集来评估机器学习模型。

11. 发布/预测结果：所有机器学习模型一旦训练好，最后都用来预测新的、未知的数据。

1.2.3　工作原理

使用TensorFlow时，必须准备样本数据集、变量、占位符和机器学习模型，然后进行模型训练，改变变量状态来提高预测结果。TensorFlow通过计算图实现上述过程。这些计算图是有向无环图，并且支持并行计算。接着TensorFlow创建损失函数，通过调整计算图中的变量来最小化损失函数。TensorFlow维护模型的计算状态，每步迭代自动计算梯度。

1.2.4　参考

❑ https://www.tensorflow.org/api_docs/python/

❏ https://www.tensorflow.org/tutorials/
❏ https://github.com/jtoy/awesome-tensorflow

1.3 声明变量和张量

TensorFlow 的主要数据结构是张量，它用张量来操作计算图。在 TensorFlow 里可以把变量或者占位符声明为张量。首先，需要知道如何创建张量。

> 张量是一个广义的向量或矩阵的数学术语。对于一维向量和二维矩阵的情况，张量是 n 维的（其中 n 可以是 1、2 甚至更大）。

1.3.1 开始

创建一个张量，声明其为一个变量。TensorFlow 在计算图中可以创建多个图结构。这里需要指出，在 TensorFlow 中创建一个张量，并不会立即在计算图中增加什么。只有运行一个操作来初始化变量之后，TensorFlow 才会把此张量增加到计算图。更多信息请见下一节对变量和占位符的讨论。

1.3.2 动手做

这里将介绍在 TensorFlow 中创建张量的主要方法：
1. 固定张量：
❏ 创建指定维度的零张量。方式如下：

zero_tsr = tf.zeros([row_dim, col_dim])

❏ 创建指定维度的全 1 张量。方式如下：

ones_tsr = tf.ones([row_dim, col_dim])

❏ 创建指定维度的常数填充的张量。方式如下：

filled_tsr = tf.fill([row_dim, col_dim], 42)

❏ 用已知常数张量创建一个张量。方式如下：

constant_tsr = tf.constant([1,2,3])

 tf.constant() 函数也能广播一个值为数组，然后模拟 tf.fill() 函数的功能，具体写法为：tf.constant(42, [row_dim, col_dim])。

2. 相似形状的张量：
❏ 新建一个与给定的 tensor 类型大小一致的 tensor，其所有元素为 0 或者 1，使用方式

如下:

```
zeros_similar = tf.zeros_like(constant_tsr)
ones_similar = tf.ones_like(constant_tsr)
```

 因为这些张量依赖前面的张量,所以初始化时需要按序进行。如果打算一次性初始化所有张量,那么程序将会报错。

3. 序列张量:

TensorFlow 可以创建指定间隔的张量。下面函数的输出跟 numpy 包中的 range() 函数和 linspace() 函数的输出相似:

```
linear_tsr = tf.linspace(start=0, stop=1, start=3)
```

返回的张量是 [0.0, 0.5, 1.0] 序列。注意,上面的函数结果中最后一个值是 stop 值。另外一个 rang() 函数的使用方式如下:

```
integer_seq_tsr = tf.range(start=6, limit=15, delta=3)
```

返回的张量是 [6, 9, 12]。注意,这个函数结果不包括 limit 值。

4. 随机张量:

❑ 下面的 tf.random_uniform() 函数生成均匀分布的随机数:

```
randunif_tsr = tf.random_uniform([row_dim, col_dim], minval=0, maxval=1)
```

注意,这个随机均匀分布从 minval(包含 minval 值)开始到 maxval(不包含 maxval 值)结束,即(minval <= x < maxval)。

❑ tf.random_normal() 函数生成正态分布的随机数:

```
randnorm_tsr = tf.random_normal([row_dim, col_dim], mean=0.0, stddev=1.0)
```

❑ tf.truncated_normal() 函数生成带有指定边界的正态分布的随机数,其正态分布的随机数位于指定均值(期望)到两个标准差之间的区间:

```
runcnorm_tsr = tf.truncated_normal([row_dim, col_dim], mean=0.0, stddev=1.0)
```

❑ 张量 / 数组的随机化。tf.random_shuffle() 和 tf, random_crop() 可以实现此功能:

```
shuffled_output = tf.random_shuffle(input_tensor)
cropped_output = tf.random_crop(input_tensor, crop_size)
```

❑ 张量的随机剪裁。tf.random_crop() 可以实现对张量指定大小的随机剪裁。在本书的后面部分,会对具有 3 通道颜色的图像(height, width, 3)进行随机剪裁。为了固定剪裁结果的一个维度,需要在相应的维度上赋其最大值:

```
cropped_image = tf.random_crop(my_image, [height/2, width/2, 3])
```

1.3.3 工作原理

一旦创建好张量，就可以通过 tf.Variable() 函数封装张量来作为变量，更多细节见下节，使用方式如下：

```
my_var = tf.Variable(tf.zeros([row_dim, col_dim]))
```

1.3.4 延伸学习

创建张量并不一定得用 TensorFlow 内建函数，可以使用 tf.convert_to_tensor() 函数将任意 numpy 数组转换为 Python 列表，或者将常量转换为一个张量。注意，tf.convert_to_tensor() 函数也可以接受张量作为输入。

1.4 使用占位符和变量

使用 TensorFlow 计算图的关键工具是占位符和变量，也请读者务必理解两者的区别，以及什么地方该用谁。

1.4.1 开始

使用数据的关键点之一是搞清楚它是占位符还是变量。变量是 TensorFlow 机器学习算法的参数，TensorFlow 维护（调整）这些变量的状态来优化机器学习算法。占位符是 TensorFlow 对象，用于表示输入输出数据的格式，允许传入指定类型和形状的数据，并依赖计算图的计算结果，比如，期望的计算结果。

1.4.2 动手做

在 TensorFlow 中，tf.Variable() 函数创建变量，过程是输入一个张量，返回一个变量。声明变量后需要初始化变量，所谓初始化就是将变量与计算图相关联。下面是创建变量并初始化的例子：

```
my_var = tf.Variable(tf.zeros([2,3]))
sess = tf.Session()
initialize_op = tf.global_variables_initializer()
sess.run(initialize_op)
```

占位符仅仅声明数据位置，用于传入数据到计算图。占位符通过会话中的 feed_dict 参数获取数据。在计算图中使用占位符时，必须在其上执行至少一个操作。在 TensorFlow 中，初始化计算图，声明一个占位符 x，定义 y 为 x 的 identity 操作。identity 操作返回占位符传入的数据本身。结果图将在下节展示，代码如下：

```
sess = tf.Session()
x = tf.placeholder(tf.float32, shape=[2,2])
y = tf.identity(x)
x_vals = np.random.rand(2,2)
```

```
sess.run(y, feed_dict={x: x_vals})
# Note that sess.run(x, feed_dict={x: x_vals}) will result in a self-
referencing error.
```

 需要注意的是 TensorFlow 不会返回一个自关联的占位符，也就是说如果运行 sess.run(x,feed_dict={x:x_vales}) 将会报错。

1.4.3 工作原理

以零张量初始化变量，其计算图如图 1-1 所示。

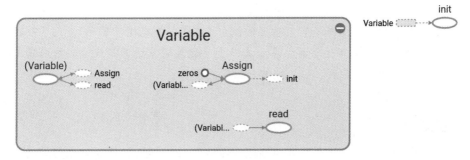

图 1-1　变量

在图 1-1 中可以看出，计算图仅仅有一个变量，全部初始化为 0。图中灰色部分详细地展示计算图操作以及相关的常量。右上角的小图展示的是主计算图。关于在 TensorFlow 中创建和可视化计算图的部分见第 10 章。

相似地，一个占位符传入 numpy 数组的计算图展示如图 1-2 所示。

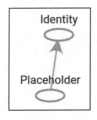

图 1-2　占位符初始化的计算图

1.4.4 延伸学习

在计算图运行的过程中，需要告诉 TensorFlow 初始化所创建的变量的时机。TensorFlow 的每个变量都有 initializer 方法，但最常用的方式是辅助函数 global_variables_initializer()。此函数会一次性初始化所创建的所有变量，使用方式如下：

```
initializer_op = tf.global_variables_initializer()
```

但是，如果是基于已经初始化的变量进行初始化，则必须按序进行初始化，方式如下：

```
sess = tf.Session()
first_var = tf.Variable(tf.zeros([2,3]))
sess.run(first_var.initializer)
second_var = tf.Variable(tf.zeros_like(first_var))
# 'second_var' depends on the 'first_var'
sess.run(second_var.initializer)
```

1.5 操作（计算）矩阵

理解 TensorFlow 如何操作矩阵，对于理解计算图中数据的流动来说非常重要。

在机器学习领域，矩阵是非常重要的概念（在数学中也同样重要）。大多数的机器学习算法均是基于矩阵的运算。鉴于本书没有涵盖矩阵运算和线性代数内容，所以建议读者能自学线性代数，以方便本书内容的理解。

1.5.1 开始

许多机器学习算法依赖矩阵操作。在 TensorFlow 中，矩阵计算是相当容易的。在下面的所有例子里，我们首先创建一个图会话，代码如下：

```
import tensorflow as tf
sess = tf.Session()
```

1.5.2 动手做

1.创建矩阵：可以使用 numpy 数组（或者嵌套列表）创建二维矩阵，也可以使用创建张量的函数（比如，zeros()、ones()、truncated_normal() 等）并为其指定一个二维形状来创建矩阵。TensorFlow 还可以使用 diag() 函数从一个一维数组（或者列表）来创建对角矩阵，代码如下：

```
identity_matrix = tf.diag([1.0, 1.0, 1.0])
A = tf.truncated_normal([2, 3])
B = tf.fill([2,3], 5.0)
C = tf.random_uniform([3,2])
D = tf.convert_to_tensor(np.array([[1., 2., 3.],[-3., -7.,
-1.],[0., 5., -2.]]))
print(sess.run(identity_matrix))
[[ 1.  0.  0.]
 [ 0.  1.  0.]
 [ 0.  0.  1.]]
print(sess.run(A))
[[ 0.96751703  0.11397751 -0.3438891 ]
 [-0.10132604 -0.8432678   0.29810596]]
print(sess.run(B))
[[ 5.  5.  5.]
```

```
 [ 5.  5.  5.]]
print(sess.run(C))
[[ 0.33184157  0.08907614]
 [ 0.53189191  0.67605299]
 [ 0.95889051  0.67061249]]
print(sess.run(D))
[[ 1.  2.  3.]
 [-3. -7. -1.]
 [ 0.  5. -2.]]
```

 注意，如果再次运行 sess.run(C)，TensorFlow 会重新初始化随机变量，并得到不同的随机数。

2. 矩阵的加法、减法和乘法：

```
print(sess.run(A+B))
[[ 4.61596632  5.39771316  4.4325695 ]
 [ 3.26702736  5.14477345  4.98265553]]
print(sess.run(B-B))
[[ 0.  0.  0.]
 [ 0.  0.  0.]]
Multiplication
print(sess.run(tf.matmul(B, identity_matrix)))
[[ 5.  5.  5.]
 [ 5.  5.  5.]]
```

矩阵乘法函数 matmul() 可以通过参数指定在矩阵乘法操作前是否进行矩阵转置或是否每个矩阵都是稀疏的。

 注意，矩阵除法没有明确定义。虽然许多人把矩阵除法定义为乘上它的倒数，但它与实数除法有本质的不同。

3. 矩阵转置，示例如下：

```
print(sess.run(tf.transpose(C)))
[[ 0.67124544  0.26766731  0.99068872]
 [ 0.25006068  0.86560275  0.58411312]]
```

再次强调，重新初始化将会得到不同的值。

4. 对于矩阵行列式，使用方式如下：

```
print(sess.run(tf.matrix_determinant(D)))
-38.0
```

5. 矩阵的逆矩阵：

```
print(sess.run(tf.matrix_inverse(D)))
[[-0.5         -0.5         -0.5        ]
 [ 0.15789474  0.05263158  0.21052632]
 [ 0.39473684  0.13157895  0.02631579]]
```

TensorFlow 中的矩阵求逆方法是 Cholesky 矩阵分解法（又称为平方根法），矩阵需要为对称正定矩阵或者可进行 LU 分解。

6. 矩阵分解：

❑ Cholesky 矩阵分解法，使用方式如下：

```
print(sess.run(tf.cholesky(identity_matrix)))
[[ 1.  0.  1.]
 [ 0.  1.  0.]
 [ 0.  0.  1.]]
```

7. 矩阵的特征值和特征向量，使用方式如下：

```
print(sess.run(tf.self_adjoint_eig(D))
[[-10.65907521  -0.22750691   2.88658212]
 [  0.21749542   0.63250104  -0.74339638]
 [  0.84526515   0.2587998    0.46749277]
 [ -0.4880805    0.73004459   0.47834331]]
```

注意，self_adjoint_eig() 函数的输出结果中，第一行为特征值，剩下的向量是对应的向量。在数学中，这种方法也称为矩阵的特征分解。

1.5.3 工作原理

TensorFlow 提供数值计算工具，并把这些计算添加到计算图中。这些部分对于简单的矩阵计算来说看似有些复杂，TensorFlow 增加这些矩阵操作到计算图进行张量计算。现在看起来这些介绍有些啰嗦，但是这有助于理解后续章节的内容。

1.6 声明操作

现在开始学习 TensorFlow 计算图的其他操作。

1.6.1 开始

除了标准数值计算外，TensorFlow 提供很多其他的操作。在使用之前，按照惯例创建一个计算图会话，代码如下：

```
import tensorflow as tf
sess = tf.Session()
```

1.6.2 动手做

TensorFlow 张量的基本操作有 add()、sub()、mul() 和 div()。注意，除特别说明外，这节所有的操作都是对张量的每个元素进行操作：

1. TensorFlow 提供 div() 函数的多种变种形式和相关的函数。

2. 值得注意的，div() 函数返回值的数据类型与输入数据类型一致。这意味着，在

Python 2 中，整数除法的实际返回是商的向下取整，即不大于商的最大整数；而 Python 3 版本中，TensorFlow 提供 truediv() 函数，其会在除法操作前强制转换整数为浮点数，所以最终的除法结果是浮点数，代码如下：

```
print(sess.run(tf.div(3, 4)))
0
print(sess.run(tf.truediv(3, 4)))
0.75
```

3. 如果要对浮点数进行整数除法，可以使用 floordiv() 函数。注意，此函数也返回浮点数结果，但是其会向下舍去小数位到最近的整数。示例如下：

```
print(sess.run(tf.floordiv(3.0,4.0)))
0.0
```

4. 另外一个重要的函数是 mod()（取模）。此函数返回除法的余数。示例如下：

```
print(sess.run(tf.mod(22.0, 5.0)))
2.0
```

5. 通过 cross() 函数计算两个张量的叉积。记住，叉积函数只为三维向量定义，所以 cross() 函数以两个三维张量作为输入，示例如下：

```
print(sess.run(tf.cross([1., 0., 0.], [0., 1., 0.])))
[ 0.  0.  1.0]
```

6. 下面给出数学函数的列表：

abs()	返回输入参数张量的绝对值
ceil()	返回输入参数张量的向上取整结果
cos()	返回输入参数张量的余弦值
exp()	返回输入参数张量的自然常数 e 的指数
floor()	返回输入参数张量的向下取整结果
inv()	返回输入参数张量的倒数
log()	返回输入参数张量的自然对数
maximum()	返回两个输入参数张量中元素的最大值
minimum()	返回两个输入参数张量中元素的最小值
neg()	返回输入参数张量的负值
pow()	返回输入参数第一个张量的第二个张量的次幂
round()	返回输入参数张量的四舍五入结果
rsqrt()	返回输入参数张量的平方根的倒数
sign()	根据输入参数张量的符号，返回 -1、0 或 1
sin()	返回输入参数张量的正弦值
sqrt()	返回输入参数张量的平方根
square()	返回输入参数张量的平方

7. 特殊数学函数：有些用在机器学习中的特殊数学函数值得一提，TensorFlow 也有对应的内建函数。除特别说明外，这些函数操作的也是张量的每个元素。

函数	说明
digamma()	普西函数（Psi 函数），lgamma() 函数的导数
erf()	返回张量的高斯误差函数
erfc()	返回张量的互补误差函数
igamma()	返回下不完全伽马函数
igammac()	返回上不完全伽马函数
lbeta()	返回贝塔函数绝对值的自然对数
lgamma()	返回伽马函数绝对值的自然对数
squared_difference()	返回两个张量间差值的平方

1.6.3　工作原理

知道在计算图中应用什么函数合适是重要的。大部分情况下，我们关心预处理函数，但也通过组合预处理函数生成许多自定义函数，示例如下：

```
# Tangent function (tan(pi/4)=1)
print(sess.run(tf.tan(3.1416/4.)))
1.0
```

1.6.4　延伸学习

如果希望在计算图中增加其他操作（未在上述函数列表中列出的操作），必须创建自定义函数。下面创建一个自定义二次多项式函数 $3x^2 - x + 10$：

```
def custom_polynomial(value):
    return tf.sub(3 * tf.square(value), value) + 10
print(sess.run(custom_polynomial(11)))
362
```

1.7　实现激励函数

1.7.1　开始

激励函数是使用所有神经网络算法的必备"神器"。激励函数的目的是为了调节权重和偏差。在 TensorFlow 中，激励函数是作用在张量上的非线性操作。激励函数的使用方法和前面的数学操作相似。激励函数的功能有很多，但其主要是为计算图归一化返回结果而引进的非线性部分。创建一个 TensorFlow 计算图：

```
import tensorflow as tf
sess = tf.Session()
```

1.7.2　动手做

TensorFlow 的激励函数位于神经网络（neural network，nn）库。除了使用 TensorFlow 内建激励函数外，我们也可以使用 TensorFlow 操作设计自定义激励函数。导入预定义激励函数（import tensorflow.nn as nn），或者在函数中显式调用 .nn。这里，选择每个函数显式调用的方法。

1. ReLU（Rectifier Linear Unit，整流线性单元）激励函数是神经网络最常用的非线性函数。其函数为 max(0, x)，连续但不平滑。示例如下：

```
print(sess.run(tf.nn.relu([-3., 3., 10.])))
[  0.   3.  10.]
```

2. 有时为了抵消 ReLU 激励函数的线性增长部分，会在 min() 函数中嵌入 max(0，x)，其在 TensorFlow 中的实现称作 ReLU6，表示为 min(max(0, x), 6)。这是 hard-sigmoid 函数的变种，计算运行速度快，解决梯度消失（无限趋近于 0），这些将在第 8 章和第 9 章中详细阐述，使用方式如下：

```
print(sess.run(tf.nn.relu6([-3., 3., 10.])))
[ 0.  3.  6.]
```

3. sigmoid 函数是最常用的连续、平滑的激励函数。它也被称作逻辑函数（Logistic 函数），表示为 1/(1+exp(-x))。sigmoid 函数由于在机器学习训练过程中反向传播项趋近于 0，因此不怎么使用。使用方式如下：

```
print(sess.run(tf.nn.sigmoid([-1., 0., 1.])))
[ 0.26894143  0.5         0.7310586 ]
```

> 注意，有些激励函数不以 0 为中心，比如，sigmoid 函数。在大部分计算图算法中要求优先使用均值为 0 的样本数据集。

4. 另外一种激励函数是双曲正切函数（tanh）。双曲正切函数与 sigmoid 函数相似，但有一点不同：双曲正切函数取值范围为 0 到 1；sigmoid 函数取值范围为 –1 到 1。双曲正切函数是双曲正弦与双曲余弦的比值，另外一种写法是 ((exp(x)-exp(-x))/(exp(x)+exp(-x))。使用方式如下：

```
print(sess.run(tf.nn.tanh([-1., 0., 1.])))
[-0.76159418  0.          0.76159418 ]
```

5. softsign 函数也是一种激励函数，表达式为：x/(abs(x) + 1)。softsign 函数是符号函数的连续（但不平滑）估计，使用方式如下：

```
print(sess.run(tf.nn.softsign([-1., 0., -1.])))
[-0.5  0.   0.5]
```

6. softplus 激励函数是 ReLU 激励函数的平滑版，表达式为：log(exp(x) + 1)。使用方式如下：

```
print(sess.run(tf.nn.softplus([-1., 0., -1.])))
[ 0.31326166  0.69314718  1.31326163]
```

> 注意，当输入增加时，softplus 激励函数趋近于无限大，softsign 函数趋近于 1；当输入减小时，softplus 激励函数趋近于 0，softsign 函数趋近于 –1。

7. ELU（Exponential Linear Unit，指数线性单元）激励函数与 softplus 激励函数相似，不同点在于：当输入无限小时，ELU 激励函数趋近于 –1，而 softplus 激励函数趋近于 0。其表达式为 (exp(x)+1) if x < 0 else x，使用方式如下：

```
print(sess.run(tf.nn.elu([-1., 0., -1.])))
[-0.63212055  0.          1.        ]
```

1.7.3　工作原理

上面这些激励函数是神经网络或其他计算图引入的非线性部分，并需要知道在什么位置使用激励函数。如果激励函数的取值范围在 0 和 1 之间，比如 sigmoid 激励函数，那计算图输出结果也只能在 0 到 1 之间取值。

如果激励函数隐藏在节点之间，就要意识到激励函数作用于传入的张量的影响。如果张量要缩放为均值为 0，就需要使用激励函数以使得尽可能多的变量在 0 附近。这暗示我们选用双曲正切（tanh）函数或者 softsign 函数，如果张量要缩放为正数，那么应当选用保留变量在正数范围内的激励函数。

1.7.4　延伸学习

图 1-3 和图 1-4 展示了不同的激励函数，从中可以看到的激励函数有 ReLU、ReLU6、softplus、ELU、sigmoid、softsign 和 tanh。

在图 1-3 中，我们可以看到四种激励函数：ReLU、ReLU6、softplus 和 ELU。这些激励函数输入值小于 0 时输出值逐渐变平，输入值大于 0 时输出值线性增长（除了 ReLU6 函数有最大值 6）。

图 1-4 展示的是 sigmoid、tanh 和 softsign 激励函数。这些激励函数都是平滑的，具有 S 型，注意有两个激励函数有水平渐近线。

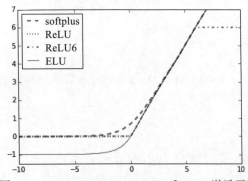

图 1-3　ReLU、ReLU6、softplus 和 ELU 激励函数

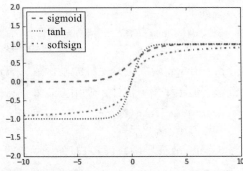

图 1-4　sigmoid、softsign 和 tanh 激励函数

1.8　读取数据源

本书中使用样本数据集训练机器学习算法模型，本节简要介绍如何通过 TensorFlow 和

Python 访问各种数据源。

 一些数据源依赖于外部网站的维护，以便你可以访问数据。如果这些网站更改或删除此数据，则可能需要更新本节中的以下某些代码。你可以在作者的 GitHub 页面上找到更新的代码：https://github.com/nfmcclure/tensorflow_cookbook。

1.8.1 开始

在 TensorFlow 中，有些数据集使用 Python 内建库，有的需要编写 Python 脚本下载，还有些得手动从网上下载。所有这些数据集都需要联网才能获取到。

1.8.2 动手做

1. 鸢尾花卉数据（Iris data）。此样本数据是机器学习和统计分析最经典的数据集，包含山鸢尾、变色鸢尾和维吉尼亚鸢尾各自的花萼和花瓣的长度和宽度。总共有 150 个数据样本，每类有 50 个样本。用 Python 加载样本数据集时，可以使用 Scikit Learn 的数据集函数，使用方式如下：

```
from sklearn import datasets
iris = datasets.load_iris()
print(len(iris.data))
150
print(len(iris.target))
150
print(iris.data[0]) # Sepal length, Sepal width, Petal length, Petal width
[ 5.1 3.5 1.4 0.2]
print(set(iris.target)) # I. setosa, I. virginica, I. versicolor
{0, 1, 2}
```

2. 出生体重数据（Birth weight data）。该数据来自 1986 年斯普林菲尔德的 Baystate 医疗中心，此样本数据集是婴儿出生体重以及母亲和家庭历史人口统计学、医学指标，有 189 个样本集，包含 11 个特征变量。使用 Python 访问数据的方式如下：

```
import requests
birthdata_url = 
'https://github.com/nfmcclure/tensorflow_cookbook/raw/master/01_Int
roduction/07_Working_with_Data_Sources/birthweight_data/birthweight
.dat'
birth_file = requests.get(birthdata_url)
birth_data = birth_file.text.split('\r\n')
birth_header = birth_data[0].split('\t')
birth_data = [[float(x) for x in y.split('\t') if len(x)>=1] for y
in birth_data[1:] if len(y)>=1]
print(len(birth_data))
189
print(len(birth_data[0]))
9
```

3. 波士顿房价数据（Boston Housing data）。此样本数据集保存在卡内基梅隆大学机器学习仓库，总共有506个房价样本，包含14个特征变量。使用Python获取数据的方式如下（通过keras库）：

```
from keras.datasets import boston_housing
(x_train, y_train), (x_test, y_test) = boston_housing.load_data()
housing_header = ['CRIM', 'ZN', 'INDUS', 'CHAS', 'NOX', 'RM',
'AGE', 'DIS', 'RAD', 'TAX', 'PTRATIO', 'B', 'LSTAT', 'MEDV']
print(x_train.shape[0])
404
print(x_train.shape[1])
13
```

4. MNIST手写体字库：MNIST手写体字库是NIST手写体字库的子样本数据集，可以网址 https://yann.lecun.com/exdb/mnist/ 下载。包含70 000张数字0到9的图片，其中60 000张标注为训练样本数据集，10 000张为测试样本数据集。TensorFlow提供内建函数来访问它，MNIST手写体字库常用来进行图像识别训练。在机器学习中，提供验证样本数据集来预防过拟合是非常重要的，TensorFlow从训练样本数据集中留出5000张图片作为验证样本数据集。这里展示使用Python访问数据的方式：

```
from tensorflow.examples.tutorials.mnist import input_data
mnist = input_data.read_data_sets("MNIST_data/"," one_hot=True)
print(len(mnist.train.images))
55000
print(len(mnist.test.images))
10000
print(len(mnist.validation.images))
5000
print(mnist.train.labels[1,:]) # The first label is a 3
[ 0.  0.  0.  1.  0.  0.  0.  0.  0.  0.]
```

5. 垃圾邮件文本数据（Spam-ham text data）。通过以下方式访问垃圾邮件文本数据：

```
import requests
import io
from zipfile import ZipFile
zip_url = 'http://archive.ics.uci.edu/ml/machine-learning-databases/00228/sms spamcollection.zip'
r = requests.get(zip_url)
z = ZipFile(io.BytesIO(r.content))
file = z.read('SMSSpamCollection')
text_data = file.decode()
text_data = text_data.encode('ascii',errors='ignore')
text_data = text_data.decode().split('\n')
text_data = [x.split('\t') for x in text_data if len(x)>=1]
[text_data_target, text_data_train] = [list(x) for x in zip(*text_data)]
print(len(text_data_train))
5574
print(set(text_data_target))
{'ham', 'spam'}
print(text_data_train[1])
Ok lar... Joking wif u oni...
```

6. 影评样本数据。此样本数据集是电影观看者的影评，分为好评和差评，可以在网站 http://www.cs.cornell.edu/people/pabo/movie-review-data/ 下载。这里用 Python 进行数据处理，使用方式如下：

```
import requests
import io
import tarfile
movie_data_url =
'http://www.cs.cornell.edu/people/pabo/movie-review-data/rt-polarit
ydata.tar.gz'
r = requests.get(movie_data_url)
# Stream data into temp object
stream_data = io.BytesIO(r.content)
tmp = io.BytesIO()
while True:
    s = stream_data.read(16384)
    if not s:
        break
    tmp.write(s)
    stream_data.close()
tmp.seek(0)
# Extract tar file
tar_file = tarfile.open(fileobj=tmp, mode="r:gz")
pos = tar_file.extractfile('rt-polaritydata/rt-polarity.pos')
neg = tar_file.extractfile('rt-polaritydata/rt-polarity.neg')
# Save pos/neg reviews (Also deal with encoding)
pos_data = []
for line in pos:
pos_data.append(line.decode('ISO-8859-1').encode('ascii',errors='ig
nore').decode())
neg_data = []
for line in neg:
neg_data.append(line.decode('ISO-8859-1').encode('ascii',errors='ig
nore').decode())
tar_file.close()
print(len(pos_data))
5331
print(len(neg_data))
5331
# Print out first negative review
print(neg_data[0])
simplistic , silly and tedious .
```

7. CIFAR-10 图像数据。此图像数据集是 CIFAR 机构发布的 8000 万张彩色图片（已缩放为 32×32 像素）的子集，总共分 10 类（包括飞机、汽车、鸟等），60 000 张图片。50 000 张图片训练数据集，10 000 张测试数据集。由于这个图像数据集数据量大，并在本书中以多种方式使用，后面到具体用时再细讲，访问网址为：http://www.cs.toronto.edu/~kriz/cifar.html。

8. 莎士比亚著作文本数据（Shakespeare text data）。此文本数据集是古登堡数字电子书计划提供的免费电子书籍，其中编译了莎士比亚的所有著作。用 Python 访问文本文件的方式如下：

```
import requests
shakespeare_url =
```

```
'http://www.gutenberg.org/cache/epub/100/pg100.txt'
# Get Shakespeare text
response = requests.get(shakespeare_url)
shakespeare_file = response.content
# Decode binary into string
shakespeare_text = shakespeare_file.decode('utf-8')
# Drop first few descriptive paragraphs.
shakespeare_text = shakespeare_text[7675:]
print(len(shakespeare_text)) # Number of characters
5582212
```

9. 英德句子翻译数据。此数据集由 Tatoeba（在线翻译数据库）发布，ManyThings.org（http://www.manythings.org）整理并提供下载。这里提供英德语句互译的文本文件（可以通过改变 URL 使用你需要的任何语言的文本文件），使用方式如下：

```
import requests
import io
from zipfile import ZipFile
sentence_url = 'http://www.manythings.org/anki/deu-eng.zip'
r = requests.get(sentence_url)
z = ZipFile(io.BytesIO(r.content))
file = z.read('deu.txt')
# Format Data
eng_ger_data = file.decode()
eng_ger_data = eng_ger_data.encode('ascii',errors='ignore')
eng_ger_data = eng_ger_data.decode().split('\n')
eng_ger_data = [x.split('\t') for x in eng_ger_data if len(x)>=1]
[english_sentence, german_sentence] = [list(x) for x in
zip(*eng_ger_data)]
print(len(english_sentence))
137673
print(len(german_sentence))
137673
print(eng_ger_data[10])
['I' won!, 'Ich habe gewonnen!']
```

1.8.3 工作原理

如果需要使用这里介绍的数据集，建议读者采用如前所述的数据下载和预处理方式。

1.8.4 参考

- Hosmer, D.W., Lemeshow, S., and Sturdivant, R. X. (2013) Applied Logistic Regression: 3rd Edition
- Lichman, M. (2013). UCI machine learning repository http://archive.ics.uci.edu/ml Irvine, CA: University of California, School of Information and Computer Science
- Bo Pang, Lillian Lee, and Shivakumar Vaithyanathan, Thumbs up? Sentiment Classification using machine learning techniques, Proceedings of EMNLP 2002 http://www.cs.cornell.edu/people/pabo/movie-review-data/
- Krizhevsky. (2009). Learning Multiple Layers of Features from Tiny Images http://www.cs.toronto.edu/~kriz/cifar.html
- Project Gutenberg. Accessed April 2016 http://www.gutenberg.org/

1.9 其他资源

这里提供一些关于 TensorFlow 使用和学习的链接、文档资料和用例。

1.9.1 开始

当开始学习使用 TensorFlow 时，需要知道在哪里能找到帮助。本节提供了一个可以帮助读者使用 TensorFlow 以及纠错的目录。

1.9.2 动手做

TensorFlow 资源列表如下：

1. 本书代码可在 GitHub（https://github.com/nfmcclure/tensorflow_cookbook）或 Packt 代码库（https://github.com/PacktPublishing/TensorFlow-Machine-Learning-Cookbook-Second-Edition）获取。

2. TensorFlow 官方 Python API 文档地址为 https://www.tensorflow.org/api_docs/python。其中包括 TensorFlow 所有函数、对象和方法的文档和例子。

3. TensorFlow 官方教程相当详细，访问网址为 https://www.tensorflow.org/tutorials/index.html。包括图像识别模型、Word2Vec、RNN 模型和 sequence-to-sequence 模型，也有些偏微分方程的例子。后续还会不断增加更多实例。

4. TensorFlow 官方 GitHub 仓库网址为 https://github.com/tensorflow/tensorflow。你可以查看源代码，甚至包含 fork 或者 clone 最新代码。也可以看到最近的 issue。

5. TensorFlow 在 Dockerhub 上维护的公开 Docker 镜像，网址为 https://hub.docker.com/r/tensorflow/tensorflow/。

6. Stack Overflow 上有 TensorFlow 标签的知识问答。随着 TensorFlow 日益流行，这个标签下的问答在不断增长，访问网址为 http://stackoverflow.com/questions/tagged/TensorFlow。

7. TensorFlow 非常灵活，应用场景广，最常用的是深度学习。为了理解深度学习的基础，数学知识和深度学习开发，Google 在在线课程 Udacity 上开课，网址为 https://www.udacity.com/course/deep-learning--ud730。

8. TensorFlow 也提供一个网站，让你可以可视化地查看随着参数和样本数据集的变化对训练神经网络的影响，网址为 http://playground.tensorflow.org。

9. 深度学习开山祖师爷 Geoffrey Hinton 在 Coursera 上开课教授"机器学习中的神经网络"，网址为 https://www.coursera.org/learn/neural-networks。

10. 斯坦福大学提供在线课程"图像识别中卷积神经网络"及其详细的课件，网址为 http://cs231n.stanford.edu/。

CHAPTER 2
第 2 章

TensorFlow 进阶

本章将介绍如何使用 TensorFlow 的关键组件，并串联起来创建一个简单的分类器，评估输出结果。阅读本章你会学到以下知识点：
- 计算图中的操作
- TensorFlow 的嵌入 Layer
- TensorFlow 的多层 Layer
- TensorFlow 实现损失函数
- TensorFlow 实现反向传播
- TensorFlow 实现随机训练和批量训练
- TensorFlow 实现创建分类器
- TensorFlow 实现模型评估

2.1 简介

现在我们已经学习完 TensorFlow 如何创建张量，使用变量和占位符；下面将把这些对象组成一个计算图。基于此，创建一个简单的分类器，并看下性能如何。

 本书的所有源代码可以在 GitHub（https://github.com/nfmcclure/tensorflow_cookbook）下载。

2.2 计算图中的操作

现在可以把这些对象表示成计算图，下面介绍计算图中作用于对象的操作。

2.2.1 开始

导入 TensorFlow，创建一个会话，开始一个计算图：

```
import tensorflow as tf
sess = tf.Session()
```

2.2.2 动手做

在这个例子中,我们将结合前面所学的知识,传入一个列表到计算图中的操作,并打印返回值:

1. 首先,声明张量和占位符。这里,创建一个 numpy 数组,传入计算图操作:

```
import numpy as np
x_vals = np.array([1., 3., 5., 7., 9.])
x_data = tf.placeholder(tf.float32)
m_const = tf.constant(3.)
my_product = tf.multiply(x_data, m_const)
for x_val in x_vals:
    print(sess.run(my_product, feed_dict={x_data: x_val}))
```

上述代码的输出如下所示:

```
3.0
9.0
15.0
21.0
27.0
```

2.2.3 工作原理

首先,创建数据集和计算图操作,然后传入数据、打印返回值。下面展示计算图(见图 2-1):

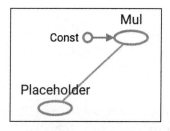

图 2-1 图中展示了占位符 x_data、乘法常量传入乘法操作

2.3 TensorFlow 的嵌入 Layer

在本节,我们将学习如何在同一个计算图中进行多个乘法操作。

2.3.1 开始

下面我们将用两个矩阵乘以占位符,然后做加法。传入两个矩阵(三维 numpy 数组):

```
import tensorflow as tf
sess = tf.Session()
```

2.3.2 动手做

知道数据在传入后是如何改变形状的也是非常重要的。我们将传入两个形状为 3×5 的 numpy 数组，然后每个矩阵乘以常量矩阵（形状为 5×1），将返回一个形状为 3×1 的矩阵。紧接着再乘以 1×1 的矩阵，返回的结果矩阵仍然为 3×1。最后，加上一个 3×1 的矩阵，示例如下：

1. 首先，创建数据和占位符：

```
my_array = np.array([[1., 3., 5., 7., 9.],
                    [-2., 0., 2., 4., 6.],
                    [-6., -3., 0., 3., 6.]])
x_vals = np.array([my_array, my_array + 1])
x_data = tf.placeholder(tf.float32, shape=(3, 5))
```

2. 接着，创建矩阵乘法和加法中要用到的常量矩阵：

```
m1 = tf.constant([[1.], [0.], [-1.], [2.], [4.]])
m2 = tf.constant([[2.]])
a1 = tf.constant([[10.]])
```

3. 现在声明操作，表示成计算图：

```
prod1 = tf.matmul(x_data, m1)
prod2 = tf.matmul(prod1, m2)
add1 = tf.add(prod2, a1)
```

4. 最后，通过计算图传入数据：

```
for x_val in x_vals:
    print(sess.run(add1, feed_dict={x_data: x_val}))
[[ 102.]
 [  66.]
 [  58.]]
[[ 114.]
 [  78.]
 [  70.]]
```

2.3.3 工作原理

上面创建的计算图可以用 Tensorboard 可视化。Tensorboard 是 TensorFlow 的功能，允许用户在图中可视化计算图和值。这些功能是原生的，不像其他机器学习框架。如果想知道这是如何做到的，可参见第 11 章。图 2-2 是分层的计算图。

2.3.4 延伸学习

在我们通过计算图运行数据之前要提前估计好声明数据的形状以及预估操作返回值的形状。由于预先不知道，或者维度在变化，情况也可能发生变化。为了实现目标，我们指明变化的维度，或者事先不知道的维度设

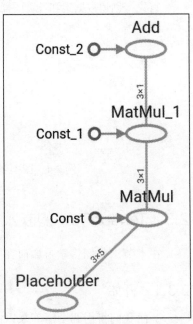

图 2-2　在图中可以看到向上传播的计算图的数据大小

为 None。例如，占位符列数未知，使用方式如下：

```
x_data = tf.placeholder(tf.float32, shape=(3,None))
```

上面虽然允许打破矩阵乘法规则，但仍然需要遵守——乘以常量矩阵返回值有一致的行数。在计算图中，也可以传入动态的 x_data，或者更改形状的 x_data，具体细节将在多批量传入数据时讲解。

 尽管可以使用 None 调节变量在某个维度上的大小，但是建议读者尽量能够明确变量的形状，并在代码中明确。None 维度主要应用在限制训练或者测试时的数据批量大小（即一次计算时多少个数据点参与运算）方面。

2.4 TensorFlow 的多层 Layer

目前，我们已经学完在同一个计算图中进行多个操作，接下来将讲述如何连接传播数据的多个层。

2.4.1 开始

本节中，将介绍如何更好地连接多层 Layer，包括自定义 Layer。这里给出一个例子（数据是生成随机图片数据），以更好地理解不同类型的操作和如何用内建层 Layer 进行计算。我们对 2D 图像进行滑动窗口平均，然后通过自定义操作层 Layer 返回结果。

在这节，我们将会看到 TensorFlow 的计算图太大，导致无法完整查看。为了解决此问题，将对各层 Layer 和操作进行层级命名管理。按照惯例，加载 numpy 和 tensorflow 模块，创建计算图，代码如下：

```
import tensorflow as tf
import numpy as np
sess = tf.Session()
```

2.4.2 动手做

1. 首先，通过 numpy 创建 2D 图像，4×4 像素图片。我们将创建成四维：第一维和最后一维大小为 1。注意，TensorFlow 的图像函数是处理四维图片的，这四维是：图片数量、高度、宽度和颜色通道。这里是一张图片，单颜色通道，所以设两个维度值为 1：

```
x_shape = [1, 4, 4, 1]
x_val = np.random.uniform(size=x_shape)
```

2. 下面在计算图中创建占位符。此例中占位符是用来传入图片的，代码如下：

```
x_data = tf.placeholder(tf.float32, shape=x_shape)
```

3. 为了创建过滤 4×4 像素图片的滑动窗口，我们将用 TensorFlow 内建函数 conv2d()

（常用来做图像处理）卷积2×2形状的常量窗口。conv2d()函数传入滑动窗口、过滤器和步长。本例将在滑动窗口四个方向上计算，所以在四个方向上都要指定步长。创建一个2×2的窗口，每个方向长度为2的步长。为了计算平均值，我们将用常量为0.25的向量与2×2的窗口卷积，代码如下：

```
my_filter = tf.constant(0.25, shape=[2, 2, 1, 1])
my_strides = [1, 2, 2, 1]
mov_avg_layer= tf.nn.conv2d(x_data, my_filter, my_strides,
                            padding='SAME',
name='Moving_Avg_Window')
```

可以使用函数中的name参数将层命名为Moring-Arg-Window。还可以使用公式：Output = (W − F + 2P)/S + 1计算卷积层的返回值形状。这里，W是输入形状，F是过滤器形状，P是padding的大小，S是步长形状。

4. 现在定义一个自定义Layer，操作滑动窗口平均的2×2的返回值。自定义函数将输入张量乘以一个2×2的矩阵张量，然后每个元素加1。因为矩阵乘法只计算二维矩阵，所以剪裁图像的多余维度（大小为1）。TensorFlow通过内建函数squeeze()剪裁。下面是新定义的Layer：

```
def custom_layer(input_matrix):
    input_matrix_sqeezed = tf.squeeze(input_matrix)
    A = tf.constant([[1., 2.], [-1., 3.]])
    b = tf.constant(1., shape=[2, 2])
    temp1 = tf.matmul(A, input_matrix_sqeezed)
    temp = tf.add(temp1, b) # Ax + b
    return tf.sigmoid(temp)
```

5. 现在把刚刚新定义的Layer加入到计算图中，并且用tf.name_scope()命名唯一的Layer名字，后续在计算图中可折叠/扩展Custom_Layer层，代码如下：

```
with tf.name_scope('Custom_Layer') as scope:
    custom_layer1 = custom_layer(mov_avg_layer)
```

6. 为占位符传入4×4像素图片，然后执行计算图，代码如下：

```
print(sess.run(custom_layer1, feed_dict={x_data: x_val}))
[[ 0.91914582 0.96025133]
 [ 0.87262219  0.9469803 ]]
```

2.4.3　工作原理

已命名的层级Layer和操作的可视化图看起来更清晰，我们可以折叠和展开已命名的自定义层Layer。在图2-3中，我们可以在左边看到折叠的概略图，在右边看到展开的详细图：

2.5　TensorFlow实现损失函数

损失函数（loss function）对机器学习来讲是非常重要的。它们度量模型输出值与目标值

（target）间的差值。本节会介绍 TensorFlow 中实现的各种损失函数。

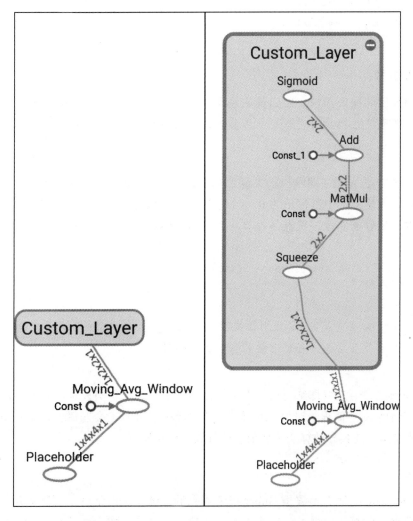

图 2-3　两层计算图。第一层是 Moving_Avg_Window，第二层是 Custom_Layer

2.5.1　开始

为了优化机器学习算法，我们需要评估机器学习模型训练输出结果。在 TensorFlow 中评估输出结果依赖损失函数。损失函数告诉 TensorFlow，预测结果相比期望的结果是好是坏。在大部分场景下，我们会有算法模型训练的样本数据集和目标值。损失函数比较预测值与目标值，并给出两者之间的数值化的差值。

本节会介绍 TensorFlow 能实现的大部分损失函数。

为了比较不同损失函数的区别，我们将会在图表中绘制出来。先创建计算图，然后加载 matplotlib（Python 的绘图库），代码如下：

```
import matplotlib.pyplot as plt
import tensorflow as tf
```

2.5.2 动手做

1. 回归算法的损失函数。回归算法是预测连续因变量的。创建预测序列和目标序列作为张量，预测序列是 −1 到 1 之间的等差数列，代码如下：

```
x_vals = tf.linspace(-1., 1., 500)
target = tf.constant(0.)
```

2. L2 正则损失函数（即欧拉损失函数）。L2 正则损失函数是预测值与目标值差值的平方和。注意，上述例子中目标值为 0。L2 正则损失函数是非常有用的损失函数，因为它在目标值附近有更好的曲度，机器学习算法利用这点收敛，并且离目标越近收敛越慢，代码如下：

```
l2_y_vals = tf.square(target - x_vals)
l2_y_out = sess.run(l2_y_vals)
```

> TensorFlow 有内建的 L2 正则形式，称为 nn.l2_loss()。这个函数其实是实际 L2 正则的一半，换句话说，它是上面 l2_y_vals 的 1/2。

3. L1 正则损失函数（即绝对值损失函数）。与 L2 正则损失函数对差值求平方不同的是，L1 正则损失函数对差值求绝对值，其优势在于当误差较大时不会变得更陡峭。L1 正则在目标值附近不平滑，这会导致算法不能很好地收敛。代码如下：

```
l1_y_vals = tf.abs(target - x_vals)
l1_y_out = sess.run(l1_y_vals)
```

4. Pseudo-Huber 损失函数是 Huber 损失函数的连续、平滑估计，试图利用 L1 和 L2 正则削减极值处的陡峭，使得目标值附近连续。它的表达式依赖参数 delta。我们将绘图来显示 delta1 = 0.25 和 delta2 = 5 的区别，代码如下：

```
delta1 = tf.constant(0.25)
phuber1_y_vals = tf.multiply(tf.square(delta1), tf.sqrt(1. +
                    tf.square((target - x_vals)/delta1)) - 1.)
phuber1_y_out = sess.run(phuber1_y_vals)
delta2 = tf.constant(5.)
phuber2_y_vals = tf.multiply(tf.square(delta2), tf.sqrt(1. +
                    tf.square((target - x_vals)/delta2)) - 1.)
phuber2_y_out = sess.run(phuber2_y_vals)
```

现在我们转向用于分类问题的损失函数。分类损失函数是用来评估预测分类结果的。通常，模型对分类的输出为一个 0～1 之间的实数。然后我们选定一个截止点（通常为

0.5），并根据输出是否高于此点进行分类。

5. 重新给 x_vals 和 target 赋值，保存返回值并在下节绘制出来，代码如下：

```
x_vals = tf.linspace(-3., 5., 500)
target = tf.constant(1.)
targets = tf.fill([500,], 1.)
```

6. Hinge 损失函数主要用来评估支持向量机算法，但有时也用来评估神经网络算法。在本例中是计算两个目标类（-1，1）之间的损失。下面的代码中，使用目标值 1，所以预测值离 1 越近，损失函数值越小：

```
hinge_y_vals = tf.maximum(0., 1. - tf.multiply(target, x_vals))
hinge_y_out = sess.run(hinge_y_vals)
```

7. 两类交叉熵损失函数（Cross-entropy loss）有时也作为逻辑损失函数。比如，当预测两类目标 0 或者 1 时，希望度量预测值到真实分类值（0 或者 1）的距离，这个距离经常是 0 到 1 之间的实数。为了度量这个距离，我们可以使用信息论中的交叉熵，代码如下：

```
xentropy_y_vals = - tf.multiply(target, tf.log(x_vals)) -
tf.multiply((1. - target), tf.log(1. - x_vals))
xentropy_y_out = sess.run(xentropy_y_vals)
```

8. Sigmoid 交叉熵损失函数（Sigmoid cross entropy loss）与上一个损失函数非常类似，有一点不同的是，它先把 x_vals 值通过 sigmoid 函数转换，再计算交叉熵损失，代码如下：

```
xentropy_sigmoid_y_vals =
tf.nn.sigmoid_cross_entropy_with_logits_v2(logits=x_vals,
labels=targets)
xentropy_sigmoid_y_out = sess.run(xentropy_sigmoid_y_vals)
```

9. 加权交叉熵损失函数（Weighted cross entropy loss）是 Sigmoid 交叉熵损失函数的加权，对正目标加权。举个例子，我们将正目标加权权重 0.5，代码如下：

```
weight = tf.constant(0.5)
xentropy_weighted_y_vals =
tf.nn.weighted_cross_entropy_with_logits(logits=x_vals,
targets=targets, pos_weight=weight)
xentropy_weighted_y_out = sess.run(xentropy_weighted_y_vals)
```

10. Softmax 交叉熵损失函数（Softmax cross-entropy loss）是作用于非归一化的输出结果，只针对单个目标分类的计算损失。通过 softmax 函数将输出结果转化成概率分布，然后计算真值概率分布的损失，代码如下：

```
unscaled_logits = tf.constant([[1., -3., 10.]])
target_dist = tf.constant([[0.1, 0.02, 0.88]])
softmax_xentropy =
tf.nn.softmax_cross_entropy_with_logits_v2(logits=unscaled_logits,
labels=target_dist)
print(sess.run(softmax_xentropy))
[ 1.16012561]
```

11. 稀疏 Softmax 交叉熵损失函数（Sparse softmax cross-entropy loss）和上一个损失函数类似，它是把目标分类为 true 的转化成 index，而 Softmax 交叉熵损失函数将目标转成概率分布。代码如下：

```
unscaled_logits = tf.constant([[1., -3., 10.]])
sparse_target_dist = tf.constant([2])
sparse_xentropy =
tf.nn.sparse_softmax_cross_entropy_with_logits(logits=unscaled_logi
ts, labels=sparse_target_dist)
print(sess.run(sparse_xentropy))
[ 0.00012564]
```

2.5.3 工作原理

这里用matplotlib绘制回归算法的损失函数（见图2-4）：

```
x_array = sess.run(x_vals)
plt.plot(x_array, l2_y_out, 'b-', label='L2 Loss')
plt.plot(x_array, l1_y_out, 'r--', label='L1 Loss')
plt.plot(x_array, phuber1_y_out, 'k-.', label='P-Huber Loss
(0.25)')
plt.plot(x_array, phuber2_y_out, 'g:', label='P-Huber Loss (5.0)')
plt.ylim(-0.2, 0.4)
plt.legend(loc='lower right', prop={'size': 11})
plt.show()
```

下面是用matplotlib绘制各种分类算法损失函数（见图2-5）：

```
x_array = sess.run(x_vals)
plt.plot(x_array, hinge_y_out, 'b-''', label='Hinge Loss''')
plt.plot(x_array, xentropy_y_out, 'r--''', label='Cross' Entropy
Loss')
plt.plot(x_array, xentropy_sigmoid_y_out, 'k-.''', label='Cross'
Entropy Sigmoid Loss')
plt.plot(x_array, xentropy_weighted_y_out, g:''', label='Weighted'
Cross Enropy Loss (x0.5)')
plt.ylim(-1.5, 3)
plt.legend(loc='lower right''', prop={'size''': 11})
plt.show()
```

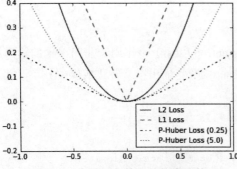

图2-4 各种回归算法的损失函数

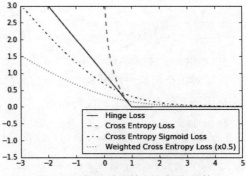

图2-5 各种分类算法的损失函数

2.5.4 延伸学习

下面总结一下前面描述的各种损失函数：

损失函数	使用类型	优 点	缺 点
L2	回归算法	更稳定	缺少健壮
L1	回归算法	更健壮	缺少稳定
Psuedo-Huber	回归算法	更健壮、稳定	参数多
Hinge	分类算法	常用于 SVM 的最大距离	异常值导致无边界损失
Cross-entropy	分类算法	更稳定	缺少健壮，出现无边界损失

其他分类算法的损失函数都需要做交叉熵损失。Sigmoid 交叉熵损失函数被用在非归一化逻辑操作，先计算 sigmoid，再计算交叉熵。TensorFlow 有很好的内建方法来处理数值边界问题。Softmax 交叉熵和稀疏 Softmax 交叉熵都类似。

这里大部分描述的分类算法损失函数是针对二类分类预测，不过也可以通过对每个预测值 / 目标的交叉熵求和，扩展成多类分类。

也有一些其他指标来评价机器学习模型，这里给出一个列表。

模型指标	描 述
R 平方值（R-squared）	对简单的线性模型来讲，用于度量因变量的变异中可由自变量解释部分所占的比例
RMSE（均方根误差）	对连续模型来讲，平均方差是度量预测的值和观察到的值之差的样本标准差
混淆矩阵（Confusion matrix）	对分类模型来讲，以矩阵形式将数据集中的记录按照真实的类别与分类模型预测的分类判断两个标准进行分析汇总，其每一列代表预测值，每一行代表的是实际的类别。理想情况下，混淆矩阵是对角矩阵
召回率（Recall）	对于分类模型来讲，召回率是正类预测为正类数与所有预测正类数的比值
精准度（Precision）	对于分类模型来讲，精准度是正类预测为正类数与所有实际正类数的比值
F 值（F-score）	对于分类模型来讲，F 值是召回率和精准度的调和平均数

2.6 TensorFlow 实现反向传播

使用 TensorFlow 的一个优势是，它可以维护操作状态和基于反向传播自动地更新模型变量。本节将介绍如何使用这种优势来训练机器学习模型。

2.6.1 开始

现在开始介绍如何调节模型变量来最小化损失函数。前面已经学习了创建对象和操作，创建度量预测值和目标值之间差值的损失函数。这里将讲解 TensorFlow 是如何通过计算图来实现最小化损失函数的误差反向传播进而更新变量的。这步将通过声明优化函数（optimization function）来实现。一旦声明好优化函数，TensorFlow 将通过它在所有的计算图中解决反向传播的项。当我们传入数据，最小化损失函数，TensorFlow 会在计算图中根据状态相应的调节变量。

本节先举个简单的回归算法的例子。从均值为 1、标准差为 0.1 的正态分布中抽样随机数，然后乘以变量 A，目标值为 10，损失函数为 L2 正则损失函数。理论上，A 的最优值是 10，因为生成的样例数据均值是 1。

第二个例子是一个简单的二值分类算法。从两个正态分布（$N(-1, 1)$ 和 $N(3, 1)$）生成 100 个数。所有从正态分布 $N(-1, 1)$ 生成的数据标为目标类 0；从正态分布 $N(3, 1)$ 生成的数据标为目标类 1，模型算法通过 sigmoid 函数将这些生成的数据转换成目标类数据。换句话讲，模型算法是 sigmoid $(x + A)$，其中，A 是要拟合的变量，理论上 $A = -1$。假设，两个正态分布的均值分别是 $m1$ 和 $m2$，则达到 A 的取值时，它们通过 $-(m1 + m2)/2$ 转换成到 0 等距的值。后面将会在 TensorFlow 中见证怎样取到相应的值。

同时，指定一个合适的学习率对机器学习算法的收敛是有帮助的。优化器类型也需要指定，前面的两个例子使用标准梯度下降法，由 TensorFlow 中的 GradientDescentOptimizer() 函数实现。

2.6.2 动手做

这里是回归算法例子：

1. 导入 Python 的数值计算模块，numpy 和 tensorflow：

```
import numpy as np
import tensorflow as tf
```

2. 创建计算图会话：

```
sess = tf.Session()
```

3. 生成数据，创建占位符和变量 A：

```
x_vals = np.random.normal(1, 0.1, 100)
y_vals = np.repeat(10., 100)
x_data = tf.placeholder(shape=[1], dtype=tf.float32)
y_target = tf.placeholder(shape=[1], dtype=tf.float32)
A = tf.Variable(tf.random_normal(shape=[1]))
```

4. 增加乘法操作：

```
my_output = tf.mul(x_data, A)
```

5. 增加 L2 正则损失函数：

```
loss = tf.square(my_output - y_target)
```

6. 现在声明变量的优化器。大部分优化器算法需要知道每步迭代的步长，这距离是由学习率控制的。如果学习率太小，机器学习算法可能耗时很长才能收敛；如果学习率太大，机器学习算法可能会跳过最优点。相应地导致梯度消失和梯度爆炸问题。学习率对算法的收敛影响较大，我们会在本节结尾探讨。在本节中使用的是标准梯度下降算法，但实际情况应该因问题而异，不同的问题使用不同的优化器算法，具体见 2.6.5 节中 Sebastian Ruder 所写的文章。

```
my_opt = tf.train.GradientDescentOptimizer(learning_rate=0.02)
train_step = my_opt.minimize(loss)
```

7. 在运行之前，需要初始化变量：

```
init = tf.global_variable_initializer()
sess.run(init)
```

 选取最优的学习率的理论很多，但真正解决机器学习算法的问题很难。2.6.5节列出了特定算法的学习率选取方法。

8. 最后一步是训练算法。我们迭代101次，并且每25次迭代打印返回结果。选择一个随机的 x 和 y，传入计算图中。TensorFlow 将自动地计算损失，调整 A 偏差来最小化损失：

```
for i in range(100):
    rand_index = np.random.choice(100)
    rand_x = [x_vals[rand_index]]
    rand_y = [y_vals[rand_index]]
    sess.run(train_step, feed_dict={x_data: rand_x, y_target: rand_y})
    if (i + 1) % 25 == 0:
        print('Step #' + str(i+1) + ' A = ' + str(sess.run(A)))
        print('Loss = ' + str(sess.run(loss, feed_dict={x_data: rand_x, y_target: rand_y})))
# Here is the output:
Step #25 A = [ 6.23402166]
Loss = 16.3173
Step #50 A = [ 8.50733757]
Loss = 3.56651
Step #75 A = [ 9.37753201]
Loss = 3.03149
Step #100 A = [ 9.80041122]
Loss = 0.0990248
```

现在将介绍简单的分类算法例子。如果先重置一下前面的 TensorFlow 计算图，我们就可以使用相同的 TensorFlow 脚本继续分类算法的例子。我们试图找到一个优化的转换方式 A，它可以把两个正态分布转换到原点，sigmoid 函数将正态分布分割成不同的两类。

9. 首先，重置计算图，并且重新初始化变量：

```
from tensorflow.python.framework import ops
ops.reset_default_graph()
sess = tf.Session()
```

10. 从正态分布（N(–1, 1), N(3, 1)）生成数据。同时也生成目标标签，占位符和偏差变量 A：

```
x_vals = np.concatenate((np.random.normal(-1, 1, 50),
np.random.normal(3, 1, 50)))
y_vals = np.concatenate((np.repeat(0., 50), np.repeat(1., 50)))
x_data = tf.placeholder(shape=[1], dtype=tf.float32)
y_target = tf.placeholder(shape=[1], dtype=tf.float32)
A = tf.Variable(tf.random_normal(mean=10, shape=[1]))
```

初始化变量 A 为 10 附近的值,远离理论值 -1。这样可以清楚地显示算法是如何从 10 收敛为 -1 的。

11. 增加转换操作。这里不必封装 sigmoid 函数,因为损失函数中会实现此功能:

```
my_output = tf.add(x_data, A)
```

12. 由于指定的损失函数期望批量数据增加一个批量数的维度,这里使用 expand_dims() 函数增加维度。下节将讨论如何使用批量变量训练,这次还是一次使用一个随机数据:

```
my_output_expanded = tf.expand_dims(my_output, 0)
y_target_expanded = tf.expand_dims(y_target, 0)
```

13. 初始化变量 A:

```
init = tf.initialize_all_variables()
sess.run(init)
```

14. 声明损失函数,这里使用一个带非归一化 logits 的交叉熵的损失函数,同时会用 sigmoid 函数转换。TensorFlow 的 nn.sigmoid_cross_entropy_with_logits() 函数实现所有这些功能,需要向它传入指定的维度,代码如下:

```
xentropy = tf.nn.sigmoid_cross_entropy_with_logits(
my_output_expanded, y_target_expanded)
```

15. 如前面回归算法的例子,增加一个优化器函数让 TensorFlow 知道如何更新和偏差变量:

```
my_opt = tf.train.GradientDescentOptimizer(0.05)
train_step = my_opt.minimize(xentropy)
```

16. 最后,通过随机选择的数据迭代几百次,相应地更新变量 A。每迭代 200 次打印出损失和变量 A 的返回值:

```
for i in range(1400):
    rand_index = np.random.choice(100)
    rand_x = [x_vals[rand_index]]
    rand_y = [y_vals[rand_index]]
    sess.run(train_step, feed_dict={x_data: rand_x, y_target: rand_y})
    if (i + 1) % 200 == 0:
        print('Step #' + str(i+1) + ' A = ' + str(sess.run(A)))
        print('Loss = ' + str(sess.run(xentropy, feed_dict={x_data: rand_x, y_target: rand_y})))
Step #200 A = [ 3.59597969]
Loss = [[ 0.00126199]]
Step #400 A = [ 0.50947344]
Loss = [[ 0.01149425]]
Step #600 A = [-0.50994617]
Loss = [[ 0.14271219]]
Step #800 A = [-0.76606178]
Loss = [[ 0.18807337]]
Step #1000 A = [-0.90859312]
Loss = [[ 0.02346182]]
Step #1200 A = [-0.86169094]
```

```
Loss = [[ 0.05427232]]
Step #1400 A = [-1.08486211]
Loss = [[ 0.04099189]]
```

2.6.3 工作原理

作为概括，总结如下几点：

1. 生成数据，所有样本均需要通过占位符进行加载。
2. 初始化占位符和变量。这两个算法中，由于使用相同的数据，所以占位符相似，同时均有一个乘法变量 A，但第二个分类算法多了一个偏差变量。
3. 创建损失函数，对于回归问题使用 L2 损失函数，对于分类问题使用交叉熵损失函数。
4. 定义一个优化器算法，所有算法均使用梯度下降。
5. 最后，通过随机数据样本进行迭代，更新变量。

2.6.4 延伸学习

前面涉及的优化器算法对学习率的选择较敏感。下面给出学习率选择总结：

学习率	优缺点	使用场景
小学习率	收敛慢，但结果精确	若算法不稳定，先降低学习率
大学习率	结果不精确，但收敛快	若算法收敛太慢，可提高学习率

有时，标准梯度下降算法会明显卡顿或者收敛变慢，特别是在梯度为 0 的附近点。为了解决此问题，TensorFlow 的 MomentumOptimizer() 函数增加了一项势能，前一次迭代过程的梯度下降值的倒数。

另外一个可以改变的是优化器的步长，理想情况下，对于变化小的变量使用大步长；而变化迅速的变量使用小步长。这里不会进行数学公式推导，但给出实现这种优点的常用算法：Adagrad 算法。此算法考虑整个历史迭代的变量梯度，TensorFlow 中相应功能的实现是 AdagradOptimizer() 函数。

有时，由于 Adagrad 算法计算整个历史迭代的梯度，导致梯度迅速变为 0。解决这个局限性的是 Adadelta 算法，它限制使用的迭代次数。TensorFlow 中相应功能的实现是 AdadeltaOptimizer() 函数。

还有一些其他的优化器算法实现，请阅读 TensorFlow 官方文档：https://www.tensorflow.org/api_guides/python/train。

2.6.5 参考

- Kingma, D., Jimmy, L. Adam: *A Method for Stochastic Optimization*. ICLR

2015 https://arxiv.org/pdf/1412.6980.pdf
- Ruder, S. *An Overview of Gradient Descent Optimization Algorithms.*
2016 https://arxiv.org/pdf/1609.04747v1.pdf
- Zeiler, M. *ADADelta: An Adaptive Learning Rate Method.*
2012 http://www.matthewzeiler.com/pubs/googleTR2012/googleTR2012.pdf

2.7 TensorFlow 实现批量训练和随机训练

根据上面描述的反向传播算法，TensorFlow 更新模型变量。它能一次操作一个数据点，也可以一次操作大量数据。一个训练例子上的操作可能导致比较"古怪"的学习过程，但使用大批量的训练会造成计算成本昂贵。到底选用哪种训练类型对机器学习算法的收敛非常关键。

2.7.1 开始

为了 TensorFlow 计算变量梯度来让反向传播工作，我们必须度量一个或者多个样本的损失。与前一节所做的相似，随机训练会一次随机抽样训练数据和目标数据对完成训练。另外一个可选项是，一次大批量训练取平均损失来进行梯度计算，批量训练大小可以一次上扩到整个数据集。这里将显示如何扩展前面的回归算法的例子——使用随机训练和批量训练。

导入 numpy、matplotlib 和 tensorflow 模块，开始一个计算图会话，代码如下：

```
import matplotlib as plt
import numpy as np
import tensorflow as tf
sess = tf.Session()
```

2.7.2 动手做

1. 开始声明批量大小。批量大小是指通过计算图一次传入多少训练数据：

```
batch_size = 20
```

2. 接下来，声明模型的数据、占位符和变量。这里能做的是改变占位符的形状，占位符有两个维度：第一个维度为 None，第二个维度是批量训练中的数据量。我们能显式地设置维度为 20，也能设为 None。如第 1 章所述，我们必须知道训练模型中的维度，这会阻止不合法的矩阵操作：

```
x_vals = np.random.normal(1, 0.1, 100)
y_vals = np.repeat(10., 100)
x_data = tf.placeholder(shape=[None, 1], dtype=tf.float32)
y_target = tf.placeholder(shape=[None, 1], dtype=tf.float32)
A = tf.Variable(tf.random_normal(shape=[1,1]))
```

3. 现在在计算图中增加矩阵乘法操作，切记矩阵乘法不满足交换律，所以在matmul()函数中的矩阵参数顺序要正确：

```
my_output = tf.matmul(x_data, A)
```

4. 改变损失函数，因为批量训练时损失函数是每个数据点L2损失的平均值。在TensorFlow中通过reduce_mean()函数即可实现，代码如下：

```
loss = tf.reduce_mean(tf.square(my_output - y_target))
```

5. 声明优化器以及初始化模型变量，代码如下：

```
my_opt = tf.train.GradientDescentOptimizer(0.02)
train_step = my_opt.minimize(loss)
init = tf.global_variables_initializer()
sess.run(init)
```

6. 在训练中通过循环迭代优化模型算法。这部分代码与之前不同，因为我们想绘制损失值图与随机训练对比，所以这里初始化一个列表每间隔5次迭代保存损失函数：

```
loss_batch = []
for i in range(100):
    rand_index = np.random.choice(100, size=batch_size)
    rand_x = np.transpose([x_vals[rand_index]])
    rand_y = np.transpose([y_vals[rand_index]])
    sess.run(train_step, feed_dict={x_data: rand_x, y_target: rand_y})
    if (i + 1) % 5 == 0:
        print('Step #' + str(i+1) + ' A = ' + str(sess.run(A)))
        temp_loss = sess.run(loss, feed_dict={x_data: rand_x, y_target: rand_y})
        print('Loss = ' + str(temp_loss))
        loss_batch.append(temp_loss)
```

7. 迭代100次输出最终返回值。注意，A值现在是二维矩阵：

```
Step #100 A = [[ 9.86720943]]
Loss = 0.
```

2.7.3 工作原理

批量训练和随机训练的不同之处在于它们的优化器方法和收敛过程。找到一个合适的批量大小是挺难的。为了展现两种训练方式收敛过程的不同，批量损失的绘图代码见下文。这里是存储随机损失的代码，接着上一节的代码：

```
loss_stochastic = []
for i in range(100):
    rand_index = np.random.choice(100)
    rand_x = [x_vals[rand_index]]
    rand_y = [y_vals[rand_index]]
    sess.run(train_step, feed_dict={x_data: rand_x, y_target: rand_y})
    if (i + 1) % 5 == 0:
        print('Step #' + str(i+1) + ' A = ' + str(sess.run(A)))
```

```
            temp_loss = sess.run(loss, feed_dict={x_data: rand_x,
y_target: rand_y})
            print('Loss = ' + str(temp_loss))
            loss_stochastic.append(temp_loss)
```

绘制回归算法的随机训练损失和批量训练损失（见图2-6），代码如下：

```
plt.plot(range(0, 100, 5), loss_stochastic, 'b-', label='Stochastic Loss')
plt.plot(range(0, 100, 5), loss_batch, 'r--', label='Batch' Loss, size=20')
plt.legend(loc='upper right', prop={'size': 11})
plt.show()
```

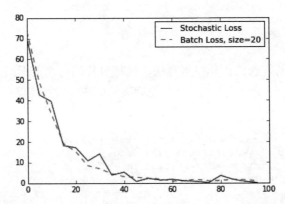

图2-6 迭代100次的随机训练损失和批量训练损失（批量大小为20）图。注意，批量训练损失更平滑，随机训练损失更不规则

2.7.4 延伸学习

训练类型	优　　点	缺　　点
随机训练	脱离局部最小	一般需更多次迭代才收敛
批量训练	快速得到最小损失	耗费更多计算资源

2.8 TensorFlow 实现创建分类器

在本节中，将结合前面所有的知识点创建一个iris数据集的分类器。

2.8.1 开始

iris数据集详细细节见第1章。加载样本数据集，实现一个简单的二值分类器来预测一朵花是否为山鸢尾。iris数据集有三类花，但这里仅预测是否是山鸢尾。导入iris数据集和工具库，相应的对原数据集进行转换。

2.8.2 动手做

1. 导入相应的工具库，初始化计算图。注意，这里导入 matplotlib 模块是为了后续绘制结果：

```
import matplotlib.pyplot as plt
import numpy as np
from sklearn import datasets
import tensorflow as tf
sess = tf.Session()
```

2. 导入 iris 数据集，根据目标数据是否为山鸢尾将其转换成 1 或者 0。由于 iris 数据集将山鸢尾标记为 0，我们将其从 0 置为 1，同时把其他物种标记为 0。本次训练只使用两种特征：花瓣长度和花瓣宽度，这两个特征在 x-value 的第三列和第四列：

```
iris = datasets.load_iris()
binary_target = np.array([1. if x==0 else 0. for x in iris.target])
iris_2d = np.array([[x[2], x[3]] for x in iris.data])
```

3. 声明批量训练大小、数据占位符和模型变量。注意，数据占位符的第一维度设为 None：

```
batch_size = 20
x1_data = tf.placeholder(shape=[None, 1], dtype=tf.float32)
x2_data = tf.placeholder(shape=[None, 1], dtype=tf.float32)
y_target = tf.placeholder(shape=[None, 1], dtype=tf.float32)
A = tf.Variable(tf.random_normal(shape=[1, 1]))
b = tf.Variable(tf.random_normal(shape=[1, 1]))
```

注意，通过指定 dtype = tf.float32 降低 float 的字节数，可以提高算法的性能。

4. 定义线性模型。线性模型的表达式为：$x2 = x1*A + b$。如果找到的数据点在直线以上，则将数据点代入 $x2 - x1*A - b$ 计算出的结果大于 0；同理找到的数据点在直线以下，则将数据点代入 $x2 - x1*A - b$ 计算出的结果小于 0。将公式 $x2 - x1*A - b$ 传入 sigmoid 函数，然后预测结果 1 或者 0。TensorFlow 有内建的 sigmoid 损失函数，所以这里仅仅需要定义模型输出即可，代码如下：

```
my_mult = tf.matmul(x2_data, A)
my_add = tf.add(my_mult, b)
my_output = tf.sub(x1_data, my_add)
```

5. 增加 TensorFlow 的 sigmoid 交叉熵损失函数 sigmoid_cross_entropy_with_logits()，代码如下：

```
xentropy = tf.nn.sigmoid_cross_entropy_with_logits(my_output, y_target)
```

6. 声明优化器方法，最小化交叉熵损失。选择学习率为 0.05，代码如下：

```
my_opt = tf.train.GradientDescentOptimizer(0.05)
train_step = my_opt.minimize(xentropy)
```

7. 创建一个变量初始化操作，然后让 TensorFlow 执行它，代码如下：

```
init = tf.global_variables_initializer()
sess.run(init)
```

8. 现在迭代 100 次训练线性模型。传入三种数据：花瓣长度、花瓣宽度和目标变量。每 200 次迭代打印出变量值，代码如下：

```
for i in range(1000):
    rand_index = np.random.choice(len(iris_2d), size=batch_size)
    rand_x = iris_2d[rand_index]
    rand_x1 = np.array([[x[0]] for x in rand_x])
    rand_x2 = np.array([[x[1]] for x in rand_x])
    rand_y = np.array([[y] for y in binary_target[rand_index]])
    sess.run(train_step, feed_dict={x1_data: rand_x1, x2_data: rand_x2, y_target: rand_y})
    if (i + 1) % 200 == 0:
        print('Step #' + str(i+1) + ' A = ' + str(sess.run(A)) + ', b = ' + str(sess.run(b)))
Step #200 A = [[ 8.67285347]], b = [[-3.47147632]]
Step #400 A = [[ 10.25393486]], b = [[-4.62928772]]
Step #600 A = [[ 11.152668]], b = [[-5.4077611]]
Step #800 A = [[ 11.81016064]], b = [[-5.96689034]]
Step #1000 A = [[ 12.41202831]], b = [[-6.34769201]]
```

9. 下面的命令抽取模型变量并绘图，结果图在下一小节展示，代码如下：

```
[[slope]] = sess.run(A)
[[intercept]] = sess.run(b)
x = np.linspace(0, 3, num=50)
ablineValues = []
for i in x:
    ablineValues.append(slope*i+intercept)

setosa_x = [a[1] for i,a in enumerate(iris_2d) if binary_target[i]==1]
setosa_y = [a[0] for i,a in enumerate(iris_2d) if binary_target[i]==1]
non_setosa_x = [a[1] for i,a in enumerate(iris_2d) if binary_target[i]==0]
non_setosa_y = [a[0] for i,a in enumerate(iris_2d) if binary_target[i]==0]
plt.plot(setosa_x, setosa_y, 'rx', ms=10, mew=2, label='setosa')
plt.plot(non_setosa_x, non_setosa_y, 'ro', label='Non-setosa')
plt.plot(x, ablineValues, 'b-')
plt.xlim([0.0, 2.7])
plt.ylim([0.0, 7.1])
plt.suptitle('Linear' Separator For I.setosa', fontsize=20)
plt.xlabel('Petal Length')
plt.ylabel('Petal Width')
plt.legend(loc='lower right')
plt.show()
```

2.8.3 工作原理

我们的目的是利用花瓣长度和花瓣宽度的特征在山鸢尾与其他物种间拟合一条直线。

绘制所有的数据点和拟合结果，将会看到图 2-7。

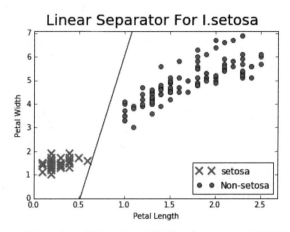

图 2-7　山鸢尾和非山鸢尾。实心直线是迭代 1000 次得到的线性分隔

2.8.4　延伸学习

当前用一条直线分割两类目标并不是最好的模型。第 4 章将会介绍一种更好的方法来分割两类目标。

2.8.5　参考

关于 iris 数据集的介绍，可以看维基百科：https://en.wikipedia.org/wiki/Iris_flower_data_set。或者 Scikit Learn 的 iris 数据集：http://scikit-learn.org/stable/auto_examples/datasets/plot_iris_dataset.html。

2.9　TensorFlow 实现模型评估

学完如何使用 TensorFlow 训练回归算法和分类算法，我们需要评估模型预测值来评估训练的好坏。

2.9.1　开始

模型评估是非常重要的，每个模型都有很多模型评估方式。使用 TensorFlow 时，需要把模型评估加入到计算图中，然后在模型训练完后调用模型评估。

在训练模型过程中，模型评估能洞察模型算法，给出提示信息来调试、提高或者改变整个模型。但是在模型训练中并不是总需要模型评估，我们将展示如何在回归算法和分类算法中使用它。

训练模型之后，需要定量评估模型的性能如何。在理想情况下，评估模型需要一个训练数据集和测试数据集，有时甚至需要一个验证数据集。

想评估一个模型时就得使用大批量数据点。如果完成批量训练，我们可以重用模型来预测批量数据点。但是如果要完成随机训练，就不得不创建单独的评估器来处理批量数据点。

 如果在损失函数中使用的模型输出结果经过转换操作，例如，sigmoid_cross_entropy_with_logits() 函数，为了精确计算预测结果，别忘了在模型评估中也要进行转换操作。

另外一个重要方面是在评估前注意是回归模型还是分类模型。

回归算法模型用来预测连续数值型，其目标不是分类值而是数字。为了评估这些回归预测值是否与实际目标相符，我们需要度量两者间的距离。这里将重写本章上一小节的回归算法的例子，打印训练过程中的损失，最终评估模型损失。

分类算法模型基于数值型输入预测分类值，实际目标是 1 和 0 的序列。我们需要度量预测值与真实值之间的距离。分类算法模型的损失函数一般不容易解释模型好坏，所以通常情况是看下准确预测分类的结果的百分比。这次将使用本章上一小节的分类算法的例子。

2.9.2 动手做

首先，将展示如何评估简单的回归算法模型，其拟合常数乘法，目标值是 10，步骤如下：

1.加载所需的编程库，创建计算图、数据集、变量和占位符。创建完数据后，将它们随机分割成训练数据集和测试数据集。不管算法模型预测的如何，我们都需要测试算法模型，这点相当重要。在训练数据和测试数据上都进行模型评估，以搞清楚模型是否过拟合：

```
import matplotlib.pyplot as plt
import numpy as np
import tensorflow as tf
sess = tf.Session()
x_vals = np.random.normal(1, 0.1, 100)
y_vals = np.repeat(10., 100)
x_data = tf.placeholder(shape=[None, 1], dtype=tf.float32)
y_target = tf.placeholder(shape=[None, 1], dtype=tf.float32)
batch_size = 25
train_indices = np.random.choice(len(x_vals),
round(len(x_vals)*0.8), replace=False)
test_indices = np.array(list(set(range(len(x_vals))) -
set(train_indices)))
x_vals_train = x_vals[train_indices]
x_vals_test = x_vals[test_indices]
y_vals_train = y_vals[train_indices]
y_vals_test = y_vals[test_indices]
A = tf.Variable(tf.random_normal(shape=[1,1]))
```

2.声明算法模型、损失函数和优化器算法。初始化模型变量 A，代码如下：

```
my_output = tf.matmul(x_data, A)
loss = tf.reduce_mean(tf.square(my_output - y_target))
my_opt = tf.train.GradientDescentOptimizer(0.02)
train_step = my_opt.minimize(loss)
init = tf.global_variables_initializer()
sess.run(init)
```

3. 像以往一样迭代训练模型，代码如下：

```
for i in range(100):
    rand_index = np.random.choice(len(x_vals_train),
size=batch_size) rand_x = np.transpose([x_vals_train[rand_index]])
    rand_y = np.transpose([y_vals_train[rand_index]])
    sess.run(train_step, feed_dict={x_data: rand_x, y_target:
rand_y})
    if (i + 1) % 25 == 0:
        print('Step #' + str(i+1) + ' A = ' + str(sess.run(A)))
        print('Loss = ' + str(sess.run(loss, feed_dict={x_data:
rand_x, y_target: rand_y})))
Step #25 A = [[ 6.39879179]]
Loss = 13.7903
Step #50 A = [[ 8.64770794]]
Loss = 2.53685
Step #75 A = [[ 9.40029907]]
Loss = 0.818259
Step #100 A = [[ 9.6809473]]
Loss = 1.10908
```

4. 现在，为了评估训练模型，将打印训练数据集和测试数据集训练的 MSE 损失函数值，代码如下：

```
mse_test = sess.run(loss, feed_dict={x_data:
np.transpose([x_vals_test]), y_target:
np.transpose([y_vals_test])})
mse_train = sess.run(loss, feed_dict={x_data:
np.transpose([x_vals_train]), y_target:
np.transpose([y_vals_train])})
print('MSE' on test:' + str(np.round(mse_test, 2)))
print('MSE' on train:' + str(np.round(mse_train, 2)))
MSE on test:1.35
MSE on train:0.88
```

对于分类模型的例子，与前面的例子类似。创建准确率函数（accuracy function），分别调用 sigmoid 来测试分类是否正确。

5. 重新加载计算图，创建数据集、变量和占位符。记住，分割数据集和目标成为训练集和测试集，代码如下：

```
from tensorflow.python.framework import ops
ops.reset_default_graph()
sess = tf.Session()
batch_size = 25
x_vals = np.concatenate((np.random.normal(-1, 1, 50),
np.random.normal(2, 1, 50)))
y_vals = np.concatenate((np.repeat(0., 50), np.repeat(1., 50)))
x_data = tf.placeholder(shape=[1, None], dtype=tf.float32)
y_target = tf.placeholder(shape=[1, None], dtype=tf.float32)
train_indices = np.random.choice(len(x_vals),
round(len(x_vals)*0.8), replace=False)
test_indices = np.array(list(set(range(len(x_vals))) -
set(train_indices)))
x_vals_train = x_vals[train_indices]
x_vals_test = x_vals[test_indices]
y_vals_train = y_vals[train_indices]
```

```
y_vals_test = y_vals[test_indices]
A = tf.Variable(tf.random_normal(mean=10, shape=[1]))
```

6. 在计算图中,增加模型和损失函数,初始化变量,并创建优化器,代码如下:

```
my_output = tf.add(x_data, A)
init = tf.initialize_all_variables()
sess.run(init)
xentropy =
tf.reduce_mean(tf.nn.sigmoid_cross_entropy_with_logits(my_output,
y_target))
my_opt = tf.train.GradientDescentOptimizer(0.05)
train_step = my_opt.minimize(xentropy)
```

7. 现在进行迭代训练,代码如下:

```
for i in range(1800):
    rand_index = np.random.choice(len(x_vals_train), size=batch_size)
    rand_x = [x_vals_train[rand_index]]
    rand_y = [y_vals_train[rand_index]]
    sess.run(train_step, feed_dict={x_data: rand_x, y_target: rand_y})
    if (i+1)%200==0:
        print('Step #' + str(i+1) + ' A = ' + str(sess.run(A)))
        print('Loss = ' + str(sess.run(xentropy, feed_dict={x_data: rand_x, y_target: rand_y})))
Step #200 A = [ 6.64970636]
Loss = 3.39434
Step #400 A = [ 2.2884655]
Loss = 0.456173
Step #600 A = [ 0.29109824]
Loss = 0.312162
Step #800 A = [-0.20045301]
Loss = 0.241349
Step #1000 A = [-0.33634067]
Loss = 0.376786
Step #1200 A = [-0.36866501]
Loss = 0.271654
Step #1400 A = [-0.3727718]
Loss = 0.294866
Step #1600 A = [-0.39153299]
Loss = 0.202275
Step #1800 A = [-0.36630616]
Loss = 0.358463
```

8. 为了评估训练模型,我们创建预测操作。用 squeeze() 函数封装预测操作,使得预测值和目标值有相同的维度。然后用 equal() 函数检测是否相等,把得到的 true 或 false 的 boolean 型张量转化成 float32 型,再对其取平均值,得到一个准确度值。我们将用这个函数评估训练模型和测试模型,代码如下:

```
y_prediction = tf.squeeze(tf.round(tf.nn.sigmoid(tf.add(x_data, A))))
correct_prediction = tf.equal(y_prediction, y_target)
accuracy = tf.reduce_mean(tf.cast(correct_prediction, tf.float32))
acc_value_test = sess.run(accuracy, feed_dict={x_data: [x_vals_test], y_target: [y_vals_test]})
```

```
acc_value_train = sess.run(accuracy, feed_dict={x_data:
[x_vals_train], y_target: [y_vals_train]})
print('Accuracy' on train set: ' + str(acc_value_train))
print('Accuracy' on test set: ' + str(acc_value_test))
Accuracy on train set: 0.925
Accuracy on test set: 0.95
```

9. 模型训练结果，比如准确度、MSE 等，将帮助我们评估机器学习模型。因为这是一维模型，能很容易地绘制模型和数据点。用 matplotlib 绘制两个分开的直方图来可视化机器学习模型和数据点（见图 2-8）：

```
A_result = sess.run(A)
bins = np.linspace(-5, 5, 50)
plt.hist(x_vals[0:50], bins, alpha=0.5, label='N(-1,1)',
color='white')
plt.hist(x_vals[50:100], bins[0:50], alpha=0.5, label='N(2,1)',
color='red')
plt.plot((A_result, A_result), (0, 8), 'k--', linewidth=3, label='A
= '+ str(np.round(A_result, 2)))
plt.legend(loc='upper right')
plt.title('Binary Classifier, Accuracy=' + str(np.round(acc_value,
2)))
plt.show()
```

2.9.3　工作原理

图中的结果标明了两类直方图的最佳分割。

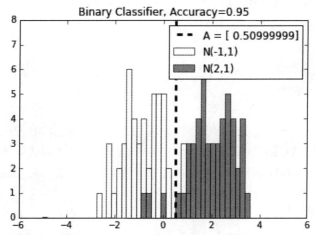

图 2-8　模型 A 和数据点的可视化。两个正态分布（均值分别为 –1 和 2），理论上的最佳分割点是 0.5，这里模型结果值（0.509 999 99）非常接近理论值 0.5

CHAPTER 3

第 3 章

基于 TensorFlow 的线性回归

本章将介绍 TensorFlow 是如何工作的，以及如何访问本书的数据集和补充学习资源。学完本章将掌握以下知识点：
- 用 TensorFlow 求逆矩阵
- 用 TensorFlow 实现矩阵分解
- 用 TensorFlow 实现线性回归
- 理解线性回归中的损失函数
- 用 TensorFlow 实现戴明回归（Deming Regression）
- 用 TensorFlow 实现 Lasso 回归和岭回归（Ridge Regression）
- 用 TensorFlow 实现弹性网络回归（Elastic Net Regression）
- 用 TensorFlow 实现逻辑回归

3.1 简介

线性回归算法是统计分析、机器学习和科学计算中最重要的算法之一，也是最常使用的算法之一，所以需要理解其是如何实现的，以及线性回归算法的各种优点。相对于许多其他算法来讲，线性回归算法是最易解释的。以每个特征的数值直接代表该特征对目标值或者因变量的影响。本章将揭晓线性回归算法的经典实现，然后讲解其在 TensorFlow 中的实现。

 请读者注意，本书的所有源码均可以在 GitHub 中访问，网址为 https://github.com/nfmcclure/tensorflow_cookbook，也可访问 Packt 代码库 https://github.com/PackPublishing/TensorFlow-Machine-Learning-Cookbook-Second-Edition。

3.2 用 TensorFlow 求逆矩阵

本节将使用 TensorFlow 求逆矩阵的方法解决二维线性回归问题。

3.2.1 开始

线性回归算法能表示为矩阵计算，$Ax = b$。这里要解决的是用矩阵 x 来求解系数。注意，如果观测矩阵不是方阵，那求解出的矩阵 x 为 $x = (A^TA)^{-1}A^Tb$。为了更直观地展示这种情况，我们将生成二维数据，用 TensorFlow 来求解，然后绘制最终结果（见图3-1）。

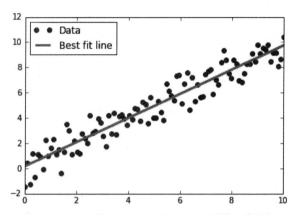

图 3-1 通过矩阵求逆方法求解拟合直线和数据点

3.2.2 动手做

1. 导入必要的编程库，初始化计算图，并生成数据，代码如下：

```
import matplotlib.pyplot as plt
import numpy as np
import tensorflow as tf
sess = tf.Session()
x_vals = np.linspace(0, 10, 100)
y_vals = x_vals + np.random.normal(0, 1, 100)
```

2. 创建后续求逆方法所需的矩阵。创建 A 矩阵，其为矩阵 x_vals_column 和 ones_column 的合并。然后以矩阵 y_vals 创建 b 矩阵，代码如下：

```
x_vals_column = np.transpose(np.matrix(x_vals))
ones_column = np.transpose(np.matrix(np.repeat(1, 100)))
A = np.column_stack((x_vals_column, ones_column))
b = np.transpose(np.matrix(y_vals))
```

3. 将 A 和 b 矩阵转换成张量，代码如下：

```
A_tensor = tf.constant(A)
b_tensor = tf.constant(b)
```

4. 现在，使用 TensorFlow 的 tf.matrix_inverse() 方法，代码如下：

```
tA_A = tf.matmul(tf.transpose(A_tensor), A_tensor)
tA_A_inv = tf.matrix_inverse(tA_A)
product = tf.matmul(tA_A_inv, tf.transpose(A_tensor))
solution = tf.matmul(product, b_tensor)
solution_eval = sess.run(solution)
```

5. 从解中抽取系数、斜率和 y 截距 y-intercept，代码如下：

```
slope = solution_eval[0][0]
y_intercept = solution_eval[1][0]
print('slope: ' + str(slope))
print('y_intercept: ' + str(y_intercept))
slope: 0.955707151739
y_intercept: 0.174366829314
best_fit = []
for i in x_vals:
    best_fit.append(slope*i+y_intercept)

plt.plot(x_vals, y_vals, 'o', label='Data')
plt.plot(x_vals, best_fit, 'r-', label='Best fit line',
linewidth=3)
plt.legend(loc='upper left')
plt.show()
```

3.2.3 工作原理

与本书的大部分章节不一样的是，这里的解决方法是通过矩阵操作直接求解结果。大部分 TensorFlow 算法是通过迭代训练实现的，利用反向传播自动更新模型变量。这里通过实现数据直接求解的方法拟合模型，仅仅是为了说明 TensorFlow 的灵活用法。

我们使用了二维数据例子来展示数据的拟合。需要注意的是，求取系数的公式（$x = (A^T A)^{-1} A^T b$）能够根据需要扩展到数据中任意数量（但对共线性问题无效）。

3.3 用 TensorFlow 实现矩阵分解

本节将用 TensorFlow 为线性回归算法实现矩阵分解。特别地，我们会使用 Cholesky 矩阵分解法，相关的函数已在 TensorFlow 中实现。

3.3.1 开始

在上一节中实现的求逆矩阵的方法在大部分情况下是低效率的，特别地，当矩阵非常大时效率更低。另外一种实现方法是矩阵分解，此方法使用 TensorFlow 内建的 Cholesky 矩阵分解法。用户对将一个矩阵分解为多个矩阵的方法感兴趣的原因是，结果矩阵的特性使得其在应用中更高效。Cholesky 矩阵分解法把一个矩阵分解为上三角矩阵和下三角矩阵，L 和 L'（L' 和 L 互为转置矩阵）。求解 $Ax = b$，改写成 $LL'x = b$。首先求解 $Ly = b$，然后求解 $L'x = y$ 得到系数矩阵 x。

3.3.2 动手做

1. 导入编程库，初始化计算图，生成数据集。接着获取矩阵 A 和 b，代码如下：

```
import matplotlib.pyplot as plt
import numpy as np
```

```
import tensorflow as tf
from tensorflow.python.framework import ops
ops.reset_default_graph()
sess = tf.Session()
x_vals = np.linspace(0, 10, 100)
y_vals = x_vals + np.random.normal(0, 1, 100)
x_vals_column = np.transpose(np.matrix(x_vals))
ones_column = np.transpose(np.matrix(np.repeat(1, 100)))
A = np.column_stack((x_vals_column, ones_column))
b = np.transpose(np.matrix(y_vals))
A_tensor = tf.constant(A)
b_tensor = tf.constant(b)
```

2. 找到方阵的 Cholesky 矩阵分解，A^TA：

```
tA_A = tf.matmul(tf.transpose(A_tensor), A_tensor)
L = tf.cholesky(tA_A)
tA_b = tf.matmul(tf.transpose(A_tensor), b)
sol1 = tf.matrix_solve(L, tA_b)
sol2 = tf.matrix_solve(tf.transpose(L), sol1)
```

注意，TensorFlow 的 cholesky() 函数仅仅返回矩阵分解的下三角矩阵，因为上三角矩阵是下三角矩阵的转置矩阵。

3. 抽取系数：

```
solution_eval = sess.run(sol2)
slope = solution_eval[0][0]
y_intercept = solution_eval[1][0]
print('slope: ' + str(slope))
print('y_intercept: ' + str(y_intercept))
slope: 0.956117676145
y_intercept: 0.136575513864
best_fit = []
for i in x_vals:
    best_fit.append(slope*i+y_intercept)
plt.plot(x_vals, y_vals, 'o', label='Data')
plt.plot(x_vals, best_fit, 'r-', label='Best fit line',
linewidth=3)
plt.legend(loc='upper left')
plt.show()
```

3.3.3 工作原理

正如你所看到的，最终求解的结果与前一节的相似。记住，通过分解矩阵的方法求解有时更高效并且数值稳定（见图 3-2）。

3.4 用 TensorFlow 实现线性回归算法

虽然使用矩阵和分解方法非常强大，但 TensorFlow 有另一种方法来求解斜率和截距。它可以通过迭代做到这一点，逐步学习将最小化损失的最佳线性回归参数。

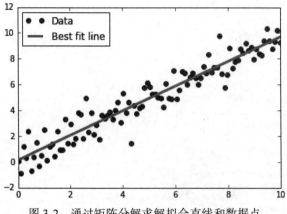

图 3-2　通过矩阵分解求解拟合直线和数据点

3.4.1　开始

本节将遍历批量数据点并让 TensorFlow 更新斜率和 y 截距。这次将使用 Scikit Learn 的内建 iris 数据集。特别地，我们将用数据点（x 值代表花瓣宽度，y 值代表花瓣长度）找到最优直线。选择这两种特征是因为它们具有线性关系，在后续结果中将会看到。下一节将讲解不同损失函数的影响，本节将使用 L2 正则损失函数。

3.4.2　动手做

1.导入必要的编程库，创建计算图，加载数据集，代码如下：

```
import matplotlib.pyplot as plt
import numpy as np
import tensorflow as tf
from sklearn import datasets
from tensorflow.python.framework import ops
ops.reset_default_graph()
sess = tf.Session()
iris = datasets.load_iris()

x_vals = np.array([x[3] for x in iris.data])
y_vals = np.array([y[0] for y in iris.data])
```

2.声明学习率、批量大小、占位符和模型变量，代码如下：

```
learning_rate = 0.05
batch_size = 25
x_data = tf.placeholder(shape=[None, 1], dtype=tf.float32)
y_target = tf.placeholder(shape=[None, 1], dtype=tf.float32)
A = tf.Variable(tf.random_normal(shape=[1,1]))
b = tf.Variable(tf.random_normal(shape=[1,1]))
```

3.增加线性模型，$y = Ax + b$，代码如下：

```
model_output = tf.add(tf.matmul(x_data, A), b)
```

4.声明 L2 损失函数，其为批量损失的平均值。初始化变量，声明优化器。注意，学习率设为 0.05，代码如下：

```
loss = tf.reduce_mean(tf.square(y_target - model_output))
init = tf.global_variables_initializer()
sess.run(init)
my_opt = tf.train.GradientDescentOptimizer(learning_rate)
train_step = my_opt.minimize(loss)
```

5. 现在遍历迭代，并在随机选择的批量数据上进行模型训练。迭代 100 次，每 25 次迭代输出变量值和损失值。注意，这里保存每次迭代的损失值，将其用于后续的可视化。代码如下：

```
loss_vec = []
for i in range(100):
    rand_index = np.random.choice(len(x_vals), size=batch_size)
    rand_x = np.transpose([x_vals[rand_index]])
    rand_y = np.transpose([y_vals[rand_index]])
    sess.run(train_step, feed_dict={x_data: rand_x, y_target: rand_y})
    temp_loss = sess.run(loss, feed_dict={x_data: rand_x, y_target: rand_y})
    loss_vec.append(temp_loss)
    if (i+1)%25==0:
        print('Step #' + str(i+1) + ' A = ' + str(sess.run(A)) + ' b = ' + str(sess.run(b)))
        print('Loss = ' + str(temp_loss))

Step #25 A = [[ 2.17270374]] b = [[ 2.85338426]]
Loss = 1.08116
Step #50 A = [[ 1.70683455]] b = [[ 3.59916329]]
Loss = 0.796941
Step #75 A = [[ 1.32762754]] b = [[ 4.08189011]]
Loss = 0.466912
Step #100 A = [[ 1.15968263]] b = [[ 4.38497639]]
Loss = 0.281003
```

6. 抽取系数，创建最佳拟合直线，代码如下：

```
[slope] = sess.run(A)
[y_intercept] = sess.run(b)
best_fit = []
for i in x_vals:
    best_fit.append(slope*i+y_intercept)
```

7. 这里将绘制两幅图。第一幅图（见图 3-3）是拟合的直线；第二幅图（见图 3-4）是迭代 100 次的 L2 正则损失函数，代码如下：

```
plt.plot(x_vals, y_vals, 'o', label='Data Points')
plt.plot(x_vals, best_fit, 'r-', label='Best fit line', linewidth=3)
plt.legend(loc='upper left')
plt.title('Sepal Length vs Petal Width')
plt.xlabel('Petal Width')
plt.ylabel('Sepal Length')
plt.show()
plt.plot(loss_vec, 'k-')
plt.title('L2 Loss per Generation')
plt.xlabel('Generation')
plt.ylabel('L2 Loss')
plt.show()
```

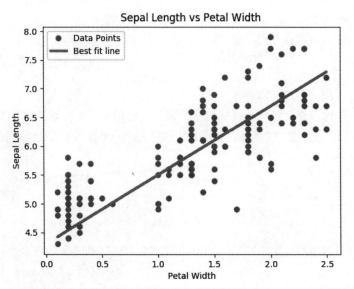

图 3-3　iris 数据集中的数据点（花瓣长度和花瓣宽度）和 TensorFlow 拟合的直线

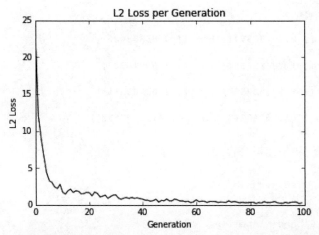

图 3-4　L2 正则损失。注意，损失函数中的抖动，批量大小越大抖动会减少；批量大小越小抖动会增大

> 这里很容易看出算法模型是过拟合还是欠拟合。将数据集分割成测试数据集和训练数据集，如果训练数据集的准确度更大，而测试数据集准确度更低，那么该拟合为过拟合；如果在测试数据集和训练数据集上的准确度都一直在增加，那么该拟合是欠拟合，需要继续训练。

3.4.3　工作原理

并不能保证最优直线是最佳拟合的直线。最佳拟合直线的收敛依赖迭代次数、批量大小、学习率和损失函数。最好时刻观察损失函数，它能帮助我们进行问题定位或者超参数调整。

3.5 理解线性回归中的损失函数

理解各种损失函数在算法收敛的影响是非常重要的。这里将展示 L1 正则和 L2 正则损失函数对线性回归算法收敛的影响。

3.5.1 开始

这次继续使用上一节中的 iris 数据集，通过改变损失函数和学习率来观察收敛性的变化。

3.5.2 动手做

1.除了损失函数外，程序的开始与以往一样，导入必要的编程库，创建一个会话，加载数据，创建占位符，定义变量和模型。我们将抽出学习率和模型迭代次数，以便展示调整这些参数的影响。代码如下：

```
import matplotlib.pyplot as plt
import numpy as np
import tensorflow as tf
from sklearn import datasets
sess = tf.Session()
iris = datasets.load_iris()
x_vals = np.array([x[3] for x in iris.data])
y_vals = np.array([y[0] for y in iris.data])
batch_size = 25
learning_rate = 0.1 # Will not converge with learning rate at 0.4
iterations = 50
x_data = tf.placeholder(shape=[None, 1], dtype=tf.float32)
y_target = tf.placeholder(shape=[None, 1], dtype=tf.float32)
A = tf.Variable(tf.random_normal(shape=[1,1]))
b = tf.Variable(tf.random_normal(shape=[1,1]))
model_output = tf.add(tf.matmul(x_data, A), b)
```

2.损失函数改为 L1 正则损失函数，代码如下：

```
loss_l1 = tf.reduce_mean(tf.abs(y_target - model_output))
```

3.现在继续初始化变量，声明优化器，遍历迭代训练。注意，为了度量收敛性，每次迭代都会保存损失值。代码如下：

```
init = tf.global_variables_initializer()
sess.run(init)
my_opt_l1 = tf.train.GradientDescentOptimizer(learning_rate)
train_step_l1 = my_opt_l1.minimize(loss_l1)
loss_vec_l1 = []
for i in range(iterations):
```

```
        rand_index = np.random.choice(len(x_vals), size=batch_size)
        rand_x = np.transpose([x_vals[rand_index]])
        rand_y = np.transpose([y_vals[rand_index]])
        sess.run(train_step_l1, feed_dict={x_data: rand_x, y_target:
rand_y})
        temp_loss_l1 = sess.run(loss_l1, feed_dict={x_data: rand_x,
y_target: rand_y})
        loss_vec_l1.append(temp_loss_l1)
        if (i+1)%25==0:
            print('Step #' + str(i+1) + ' A = ' + str(sess.run(A)) + '
b = ' + str(sess.run(b)))

plt.plot(loss_vec_l1, 'k-', label='L1 Loss')
plt.plot(loss_vec_l2, 'r--', label='L2 Loss')
plt.title('L1 and L2 Loss per Generation')
plt.xlabel('Generation')
plt.ylabel('L1 Loss')
plt.legend(loc='upper right')
plt.show()
```

3.5.3 工作原理

当选择了一个损失函数时，也要选择对应的学习率。这里展示了两种解决方法，一种是上一节的 L2 正则损失函数，另一种是 L1 正则损失函数。

如果学习率太小，算法收敛耗时将更长。但是如果学习率太大，算法有可能产生不收敛的问题。下面绘制 iris 数据的线性回归问题的 L1 正则和 L2 正则损失（见图 3-5），其中学习率为 0.05。

从图 3-5 中可以看出，当学习率为 0.05 时，L2 正则损失更优，其有更低的损失值。当学习率增加为 0.4 时，绘制其损失函数（见图 3-6）。

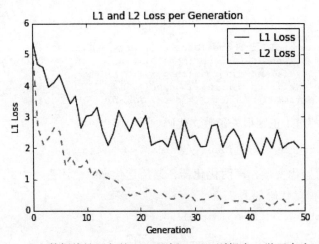

图 3-5　iris 数据线性回归的 L1 正则和 L2 正则损失，学习率为 0.05

从图 3-6 中可以发现，学习率大导致 L2 损失过大，而 L1 正则损失收敛。

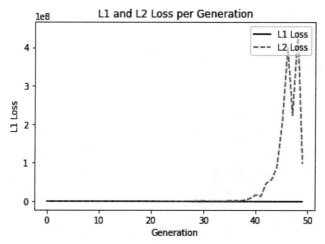

图 3-6 iris 数据线性回归的 L1 正则和 L2 正则损失，学习率为 0.4。其中 L1 正则损失不可见是因为它的 y 轴值太大

3.5.4 延伸学习

为了更容易地理解上述的情况，这里清晰地展示大学习率和小学习率对 L1 正则和 L2 正则损失函数的影响。这里可视化的是 L1 正则和 L2 正则损失函数的一维情况，如图 3-7 所示。

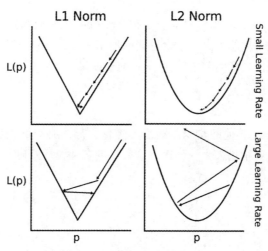

图 3-7 L1 正则和 L2 正则损失函数在大学习率和小学习率下的影响

3.6 用 TensorFlow 实现戴明回归算法

本节将实现戴明回归（Deming Regression），其意味着需要不同的方式来度量模型直线

和数据集的数据点间的距离。

 戴明回归有很多别名,例如全回归、正交回归(ODR)或者最短路径回归。

3.6.1 开始

如果最小二乘线性回归算法最小化到回归直线的竖直距离(即,平行于 y 轴方向),则戴明回归最小化到回归直线的总距离(即,垂直于回归直线)。其最小化 x 值和 y 值两个方向的误差,具体的对比图如图 3-8 所示。

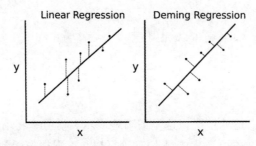

图 3-8 线性回归算法和戴明回归算法的区别。左边的线性回归最小化到回归直线的竖直距离;右边的戴明回归最小化到回归直线的总距离

为了实现戴明回归算法,我们修改一下损失函数。线性回归算法的损失函数最小化竖直距离;而这里需要最小化总距离。给定直线的斜率和截距,则求解一个点到直线的垂直距离有已知的几何公式。代入几何公式并使 TensorFlow 最小化距离。

3.6.2 动手做

1.除了损失函数外,其他的步骤跟前面的类似。导入必要的编程库,创建一个计算图会话,加载数据集,声明批量大小,创建占位符、变量和模型输出,代码如下:

```
import matplotlib.pyplot as plt
import numpy as np
import tensorflow as tf
from sklearn import datasets
sess = tf.Session()
iris = datasets.load_iris()
x_vals = np.array([x[3] for x in iris.data])
y_vals = np.array([y[0] for y in iris.data])
batch_size = 50
x_data = tf.placeholder(shape=[None, 1], dtype=tf.float32)
y_target = tf.placeholder(shape=[None, 1], dtype=tf.float32)
A = tf.Variable(tf.random_normal(shape=[1,1]))
b = tf.Variable(tf.random_normal(shape=[1,1]))
model_output = tf.add(tf.matmul(x_data, A), b)
```

2. 损失函数是由分子和分母组成的几何公式。给定直线 $y = mx + b$，点 (x_0, y_0)，则求两者间的距离的公式为：

$$d = \frac{|y_0 - (mx_0 + b)|}{\sqrt{m^2 + 1}}$$

```
deming_numerator = tf.abs(tf.sub(y_target, tf.add(tf.matmul(x_data,
A), b)))
deming_denominator = tf.sqrt(tf.add(tf.square(A),1))
loss = tf.reduce_mean(tf.truediv(deming_numerator,
deming_denominator))
```

3. 现在初始化变量，声明优化器，遍历迭代训练集以得到参数，代码如下：

```
init = tf.global_variables_initializer()
sess.run(init)
my_opt = tf.train.GradientDescentOptimizer(0.1)
train_step = my_opt.minimize(loss)
loss_vec = []
for i in range(250):
    rand_index = np.random.choice(len(x_vals), size=batch_size)
    rand_x = np.transpose([x_vals[rand_index]])
    rand_y = np.transpose([y_vals[rand_index]])
    sess.run(train_step, feed_dict={x_data: rand_x, y_target: rand_y})
    temp_loss = sess.run(loss, feed_dict={x_data: rand_x, y_target: rand_y})
    loss_vec.append(temp_loss)
    if (i+1)%50==0:
        print('Step #' + str(i+1) + ' A = ' + str(sess.run(A)) + ' b = ' + str(sess.run(b)))
        print('Loss = ' + str(temp_loss))
```

4. 绘制输出结果（见图 3-9）的代码如下：

```
[slope] = sess.run(A)
[y_intercept] = sess.run(b)
best_fit = []
for i in x_vals:
    best_fit.append(slope*i+y_intercept)

plt.plot(x_vals, y_vals, 'o', label='Data Points')
plt.plot(x_vals, best_fit, 'r-', label='Best fit line',
linewidth=3)
plt.legend(loc='upper left')
plt.title('Sepal Length vs petal Width')
plt.xlabel('petal Width')
plt.ylabel('Sepal Length')
plt.show()
```

3.6.3 工作原理

戴明回归算法与线性回归算法得到的结果基本一致。两者之间的关键不同点在于预测值与数据点间的损失函数度量：线性回归算法的损失函数是竖直距离损失；而戴明回归算法是垂直距离损失（到 x 轴和 y 轴的总距离损失）。

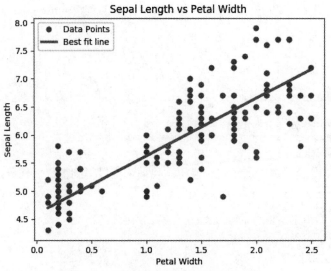

图 3-9 iris 数据集上戴明回归算法的解

> 注意，这里戴明回归算法的实现类型是总体回归（总的最小二乘法误差）。总体回归算法是假设 x 值和 y 值的误差是相似的。我们也可以根据不同的理念使用不同的误差来扩展 x 轴和 y 轴的距离计算。

3.7 用 TensorFlow 实现 lasso 回归和岭回归算法

也有些正则方法可以限制回归算法输出结果中系数的影响，其中最常用的两种正则方法是 lasso 回归和岭回归。本节将详细介绍如何实现这两种方法。

3.7.1 开始

lasso 回归和岭回归算法跟常规线性回归算法极其相似，有一点不同的是，在公式中增加正则项来限制斜率（或者净斜率）。这样做的主要原因是限制特征对因变量的影响，通过增加一个依赖斜率 A 的损失函数实现。

对于 lasso 回归算法，在损失函数上增加一项：斜率 A 的某个给定倍数。我们使用 TensorFlow 的逻辑操作，但没有这些操作相关的梯度，而是使用阶跃函数的连续估计，也称作连续阶跃函数，其会在截止点跳跃扩大。一会就可以看到如何使用 lasso 回归算法。

对于岭回归算法，增加一个 L2 范数，即斜率系数的 L2 正则。这个简单的修改将在 3.7.4 节介绍。

3.7.2 动手做

1. 这次还是使用 iris 数据集，使用方式跟前面的类似。首先，导入必要的编程库，创建一个计算图会话，加载数据集，声明批量大小，创建占位符、变量和模型输出，代码如下：

```
import matplotlib.pyplot as plt
import numpy as np
import tensorflow as tf
from sklearn import datasets
from tensorflow.python.framework import ops
ops.reset_default_graph()
sess = tf.Session()
iris = datasets.load_iris()
x_vals = np.array([x[3] for x in iris.data])
y_vals = np.array([y[0] for y in iris.data])
batch_size = 50
learning_rate = 0.001
x_data = tf.placeholder(shape=[None, 1], dtype=tf.float32)
y_target = tf.placeholder(shape=[None, 1], dtype=tf.float32)
A = tf.Variable(tf.random_normal(shape=[1,1]))
b = tf.Variable(tf.random_normal(shape=[1,1]))
model_output = tf.add(tf.matmul(x_data, A), b)
```

2. 增加损失函数，其为改良过的连续阶跃函数，lasso 回归的截止点设为 0.9。这意味着限制斜率系数不超过 0.9，代码如下：

```
lasso_param = tf.constant(0.9)
heavyside_step = tf.truediv(1., tf.add(1.,
tf.exp(tf.multiply(-100., tf.subtract(A, lasso_param)))))
regularization_param = tf.mul(heavyside_step, 99.)
loss = tf.add(tf.reduce_mean(tf.square(y_target - model_output)),
regularization_param)
```

3. 初始化变量和声明优化器，代码如下：

```
init = tf.global_variables_initializer()
sess.run(init)
my_opt = tf.train.GradientDescentOptimizer(learning_rate)
train_step = my_opt.minimize(loss)
```

4. 遍历迭代运行一段时间，因为需要过一会儿才会收敛。最后结果显示斜率系数小于 0.9，代码如下：

```
loss_vec = []
for i in range(1500):
    rand_index = np.random.choice(len(x_vals), size=batch_size)
    rand_x = np.transpose([x_vals[rand_index]])
    rand_y = np.transpose([y_vals[rand_index]])
    sess.run(train_step, feed_dict={x_data: rand_x, y_target: rand_y})
    temp_loss = sess.run(loss, feed_dict={x_data: rand_x, y_target: rand_y})
    loss_vec.append(temp_loss[0])
    if (i+1)%300==0:
        print('Step #' + str(i+1) + ' A = ' + str(sess.run(A)) + ' b = ' + str(sess.run(b)))
        print('Loss = ' + str(temp_loss))

Step #300 A = [[ 0.82512331]] b = [[ 2.30319238]]
Loss = [[ 6.84168959]]
```

```
Step #600 A = [[ 0.8200165]] b = [[ 3.45292258]]
Loss = [[ 2.02759886]]
Step #900 A = [[ 0.81428504]] b = [[ 4.08901262]]
Loss = [[ 0.49081498]]
Step #1200 A = [[ 0.80919558]] b = [[ 4.43668795]]
Loss = [[ 0.40478843]]
Step #1500 A = [[ 0.80433637]] b = [[ 4.6360755]]
Loss = [[ 0.23839757]]
```

3.7.3 工作原理

通过在标准线性回归估计的基础上，增加一个连续的阶跃函数，实现lasso回归算法。由于阶跃函数的坡度，我们需要注意步长，因为太大的步长会导致最终不收敛。对于岭回归算法，将在下一节介绍对其的必要修改。

3.7.4 延伸学习

对于岭回归算法，在上一节的代码基础上稍微改变损失函数即可，代码如下：

```
ridge_param = tf.constant(1.)
ridge_loss = tf.reduce_mean(tf.square(A))
loss = tf.expand_dims(tf.add(tf.reduce_mean(tf.square(y_target -
 model_output)), tf.multiply(ridge_param, ridge_loss)), 0)
```

3.8 用TensorFlow实现弹性网络回归算法

弹性网络回归算法（Elastic Net Regression）是综合lasso回归和岭回归的一种回归算法，通过在损失函数中增加L1和L2正则项。

3.8.1 开始

在学完前面两节之后，可以轻松地实现弹性网络回归算法。本节使用多线性回归的方法实现弹性网络回归算法，以iris数据集为训练数据，用花瓣长度、花瓣宽度和花萼宽度三个特征预测花萼长度。

3.8.2 动手做

1.导入必要的编程库并初始化一个计算图，代码如下：

```
import matplotlib.pyplot as plt
import numpy as np
import tensorflow as tf
from sklearn import datasets
sess = tf.Session()
```

2.加载数据集。这次，x_vals数据将是三列值的数组，代码如下：

```
iris = datasets.load_iris()
x_vals = np.array([[x[1], x[2], x[3]] for x in iris.data])
```

```
y_vals = np.array([y[0] for y in iris.data])
```

3. 声明批量大小、占位符、变量和模型输出。这里唯一不同的是 x_data 占位符的大小为 3，代码如下：

```
batch_size = 50
learning_rate = 0.001
x_data = tf.placeholder(shape=[None, 3], dtype=tf.float32)
y_target = tf.placeholder(shape=[None, 1], dtype=tf.float32)
A = tf.Variable(tf.random_normal(shape=[3,1]))
b = tf.Variable(tf.random_normal(shape=[1,1]))
model_output = tf.add(tf.matmul(x_data, A), b)
```

4. 对于弹性网络回归算法，损失函数包含斜率的 L1 正则和 L2 正则。创建 L1 和 L2 正则项，然后加入到损失函数中，代码如下：

```
elastic_param1 = tf.constant(1.)
elastic_param2 = tf.constant(1.)
l1_a_loss = tf.reduce_mean(tf.abs(A))
l2_a_loss = tf.reduce_mean(tf.square(A))
e1_term = tf.multiply(elastic_param1, l1_a_loss)
e2_term = tf.multiply(elastic_param2, l2_a_loss)
loss = tf.expand_dims(tf.add(tf.add(tf.reduce_mean(tf.square(y_target - model_output)), e1_term), e2_term), 0)
```

5. 现在初始化变量，声明优化器，然后遍历迭代运行，训练拟合得到系数，代码如下：

```
init = tf.global_variables_initializer()
sess.run(init)
my_opt = tf.train.GradientDescentOptimizer(learning_rate)
train_step = my_opt.minimize(loss)
loss_vec = []
for i in range(1000):
    rand_index = np.random.choice(len(x_vals), size=batch_size)
    rand_x = x_vals[rand_index]
    rand_y = np.transpose([y_vals[rand_index]])
    sess.run(train_step, feed_dict={x_data: rand_x, y_target: rand_y})
    temp_loss = sess.run(loss, feed_dict={x_data: rand_x, y_target: rand_y})
    loss_vec.append(temp_loss[0])
    if (i+1)%250==0:
        print('Step #' + str(i+1) + ' A = ' + str(sess.run(A)) + ' b = ' + str(sess.run(b)))
        print('Loss = ' + str(temp_loss))
```

6. 下面是代码运行的输出结果：

```
Step #250 A = [[ 0.42095602]
 [ 0.1055888 ]
 [ 1.77064979]] b = [[ 1.76164341]]
Loss = [ 2.87764359]
Step #500 A = [[ 0.62762028]
 [ 0.06065864]
 [ 1.36294949]] b = [[ 1.87629771]]
Loss = [ 1.8032167]
```

```
Step #750 A = [[ 0.67953539]
 [ 0.102514 ]
 [ 1.06914485]] b = [[ 1.95604002]]
Loss = [ 1.33256555]
Step #1000 A = [[ 0.6777274 ]
 [ 0.16535147]
 [ 0.8403284 ]] b = [[ 2.02246833]]
Loss = [ 1.21458709]
```

7. 现在能观察到，随着训练迭代后损失函数已收敛（见图3-10），代码如下：

```
plt.plot(loss_vec, 'k-')
plt.title('Loss per Generation')
plt.xlabel('Generation')
plt.ylabel('Loss')
plt.show()
```

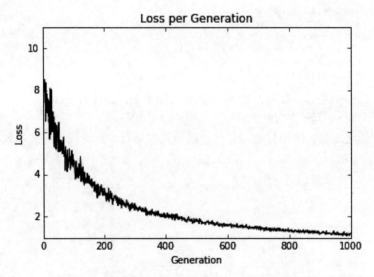

图 3-10　弹性网络回归迭代训练 1000 次的损失图

3.8.3　工作原理

弹性网络回归算法的实现是多线性回归。我们能发现，增加 L1 和 L2 正则项后的损失函数中的收敛变慢。

3.9　用 TensorFlow 实现逻辑回归算法

本节将实现逻辑回归算法，预测低出生体重的概率。

3.9.1　开始

逻辑回归算法可以将线性回归转换成一个二值分类器。通过 sigmoid 函数将线性回归的输出缩放到 0 和 1 之间。目标值是 0 或者 1 代表着一个数据点是否属于某一类。如果预测值在截止值以上，则预测值被标记为 "1" 类；否则，预测值标为 "0" 类。在本例中，为方便简单起见，将指定截止值设为 0.5。

在本例中使用的低出生体重的数据来自本书作者的 GitHub 数据仓库（https://github.com/nfmcclure/tensorflow_cookbook/raw/master/01_Introduction/07_Working_with_Data_Sources/birthweight_data/birthweight.dat）。我们将从多个因素来预测低出生体重。

3.9.2 动手做

1. 导入必要的编程库，包括 requests 模块，因为我们将通过超链接访问低出生体重数据集。初始化一个计算图，代码如下：

```
import matplotlib.pyplot as plt
import numpy as np
import tensorflow as tf
import requests
from sklearn import datasets
from sklearn.preprocessing import normalize
from tensorflow.python.framework import ops
ops.reset_default_graph()
sess = tf.Session()
```

2. 通过 requests 模块加载数据集，指定要使用的特征。实际出生体重特征和 ID 两列不需要，代码如下：

```
birth_weight_file = 'birth_weight.csv'
# Download data and create data file if file does not exist in current directory
if not os.path.exists(birth_weight_file):
    birthdata_url = 'https://github.com/nfmcclure/tensorflow_cookbook/raw/master/01_Introduction/07_Working_with_Data_Sources/birthweight_data/birthweight.dat'
    birth_file = requests.get(birthdata_url)
    birth_data = birth_file.text.split('\r\n')
    birth_header = birth_data[0].split('\t')
    birth_data = [[float(x) for x in y.split('\t') if len(x)>=1] for y in birth_data[1:] if len(y)>=1]
    with open(birth_weight_file, 'w', newline='') as f:
        writer = csv.writer(f)
        writer.writerow(birth_header)
        writer.writerows(birth_data)

# Read birth weight data into memory
birth_data = []
with open(birth_weight_file, newline='') as csvfile:
    csv_reader = csv.reader(csvfile)
    birth_header = next(csv_reader)
    for row in csv_reader:
        birth_data.append(row)
    birth_data = [[float(x) for x in row] for row in birth_data]
```

```
# Pull out target variable
y_vals = np.array([x[0] for x in birth_data])
# Pull out predictor variables (not id, not target, and not
birthweight)
x_vals = np.array([x[1:8] for x in birth_data])
```

3. 分割数据集为测试集和训练集：

```
train_indices = np.random.choice(len(x_vals),
round(len(x_vals)*0.8), replace=False)
test_indices = np.array(list(set(range(len(x_vals))) -
set(train_indices)))
x_vals_train = x_vals[train_indices]
x_vals_test = x_vals[test_indices]
y_vals_train = y_vals[train_indices]
y_vals_test = y_vals[test_indices]
```

4. 将所有特征缩放到 0 和 1 区间（min-max 缩放），逻辑回归收敛的效果更好。下面将归一化特征，代码如下：

```
def normalize_cols(m, col_min=np.array([None]),
col_max=np.array([None])):
    if not col_min[0]:
        col_min = m.min(axis=0)
    if not col_max[0]:
        col_max = m.max(axis=0)
    return (m-col_min) / (col_max - col_min), col_min, col_max
x_vals_train, train_min, train_max =
np.nan_to_num(normalize_cols(x_vals_train))
x_vals_test = np.nan_to_num(normalize_cols(x_vals_test, train_min,
train_max))
```

> 注意，在缩放数据集前，先分割数据集为测试集和训练集，这是相当重要的。我们要确保训练集和测试集互不影响。如果我们在分割数据集前先缩放，就无法保证它们不相互影响。

5. 声明批量大小、占位符、变量和逻辑模型。这步不需要用 sigmoid 函数封装输出结果，因为 sigmoid 操作是包含在内建损失函数中的，代码如下：

```
batch_size = 25
x_data = tf.placeholder(shape=[None, 7], dtype=tf.float32)
y_target = tf.placeholder(shape=[None, 1], dtype=tf.float32)
A = tf.Variable(tf.random_normal(shape=[7,1]))
b = tf.Variable(tf.random_normal(shape=[1,1]))
model_output = tf.add(tf.matmul(x_data, A), b)
```

6. 声明损失函数，其包含 sigmoid 函数。初始化变量，声明优化器，代码如下：

```
loss =
tf.reduce_mean(tf.nn.sigmoid_cross_entropy_with_logits(model_output
, y_target))
init = tf.global_variables_initializer()
```

```
sess.run(init)
my_opt = tf.train.GradientDescentOptimizer(0.01)
train_step = my_opt.minimize(loss)
```

7. 除记录损失函数外，也需要记录分类器在训练集和测试集上的准确度。所以创建一个返回准确度的预测函数，代码如下：

```
prediction = tf.round(tf.sigmoid(model_output))
predictions_correct = tf.cast(tf.equal(prediction, y_target),
tf.float32)
accuracy = tf.reduce_mean(predictions_correct)
```

8. 开始遍历迭代训练，记录损失值和准确度，代码如下：

```
loss_vec = []
train_acc = []
test_acc = []
for i in range(1500):
    rand_index = np.random.choice(len(x_vals_train),
size=batch_size)
    rand_x = x_vals_train[rand_index]
    rand_y = np.transpose([y_vals_train[rand_index]])
    sess.run(train_step, feed_dict={x_data: rand_x, y_target:
rand_y})
    temp_loss = sess.run(loss, feed_dict={x_data: rand_x, y_target:
rand_y})
    loss_vec.append(temp_loss)
    temp_acc_train = sess.run(accuracy, feed_dict={x_data:
x_vals_train, y_target: np.transpose([y_vals_train])})
    train_acc.append(temp_acc_train)
    temp_acc_test = sess.run(accuracy, feed_dict={x_data:
x_vals_test, y_target: np.transpose([y_vals_test])})
    test_acc.append(temp_acc_test)
```

9. 绘制损失和准确度，代码如下：

```
plt.plot(loss_vec, 'k-')
plt.title('Cross' Entropy Loss per Generation')
plt.xlabel('Generation')
plt.ylabel('Cross' Entropy Loss')
plt.show()
plt.plot(train_acc, 'k-', label='Train Set Accuracy')
plt.plot(test_acc, 'r--', label='Test Set Accuracy')
plt.title('Train and Test Accuracy')
plt.xlabel('Generation')
plt.ylabel('Accuracy')
plt.legend(loc='lower right')
plt.show()
```

3.9.3 工作原理

这里是迭代过程中的损失，以及训练集和测试集的准确度。数据集只有189个观测值，但训练集和测试集的准确度图由于数据集的随机分割将会变化，如图3-11和图3-12所示。

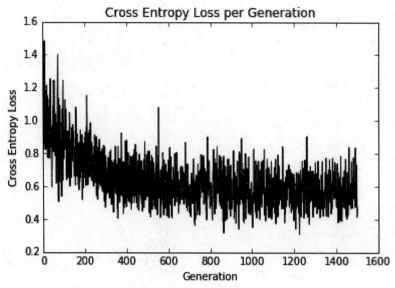

图 3-11　迭代 1500 次的交叉熵损失图

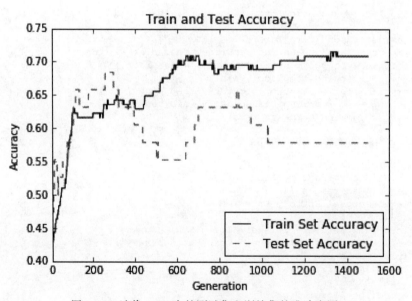

图 3-12　迭代 1500 次的测试集和训练集的准确度图

CHAPTER 4

第 4 章

基于 TensorFlow 的支持向量机

本章将详细介绍 TensorFlow 中支持向量机算法的使用、实现和评估。学完本章将掌握以下知识点：
- 线性支持向量机的使用
- 弱化为线性回归
- TensorFlow 上核函数的使用
- TensorFlow 实现非线性支持向量机
- TensorFlow 实现多类支持向量机

注意，上一章介绍的逻辑回归算法和本章的大部分支持向量机算法都是二值预测。逻辑回归算法试图找到回归直线来最大化距离（概率）；而支持向量机算法也试图最小化误差，最大化两类之间的间隔。一般来说，如果一个问题的训练集中有大量特征，则建议用逻辑回归或者线性支持向量机算法；如果训练集的数量更大，或者数据集是非线性可分的，则建议使用带高斯核的支持向量机算法。

记住，本章的所有代码在 GitHub（https://github.com/nfmcclure/tensorflow_cookbook）和 Packt 代码库（https://github.com/PacktPublishing/TensorFlow-Machine-Learning-Cookbook-Second-Edition）可以获取。

4.1 简介

支持向量机算法是一种二值分类器方法（见图 4-1）。基本的观点是，找到两类之间的一个线性可分的直线（或者超平面）。首先假设二分类目标是 –1 或者 1，代替前面章节中的 0 或者 1 目标值。有许多条直线可以分割两类目标，但是我们定义分割两类目标有最大距离的直线为最佳线性分类器。

得到一个超平面，公式如下：

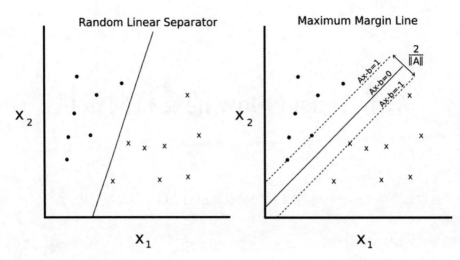

图4-1 假设两类分离的目标,"o"和"x",找到两类目标间的线性分类器等式。左边显示有许多条线分离两类目标;右边显示最大间隔的直线,间隔宽度为 $\dfrac{2}{\|A\|}$,此时为最小化系数 A 的 L2 正则因子范数

其中,A 是斜率向量,x 是输入向量。最大间隔的宽度为 2 除以 A 的 L2 范数。对于这个公式有很多种证明方式,但是,从几何观点来看,这是二维数据点到一条直线的垂直距离。

对于线性可分的二值分类数据集,为了最大化间隔,我们最小化 A 的 L2 范数 $\|A\|$。最小化也必须服从以下约束:

$$y_i(Ax_i - b) \geqslant 1 \forall i$$

上述的约束确保所有相关分类的数据点都在分割线的同一侧。

因为不是所有的数据集都是线性可分的,我们引入跨分割线的数据点的损失函数函数。对于 n 个数据点,引入 soft margin 损失函数,公式如下:

$$\frac{1}{n}\sum_{i=1}^{n}\max(0, 1 - y_i(Ax_i - b)) + a\|A\|^2$$

注意,如果数据点分割正确,乘积 $y_i(Ax_i - b)$ 总是大于 1。这意味着损失函数左边项等于 0,这时对损失函数有影响的仅仅只有间隔大小。

上述损失函数寻求一个线性可分的直线,但是也允许有些点跨越间隔直线,这取决于 a 值,当 a 值很大,模型会倾向于尽量将样本分割开;a 值越小,会有更多的跨越边界的点存在。

在本章中,将建立一个 soft margin 支持向量机,展示如何将其扩展应用到非线性的场景和多分类目标。

4.2 线性支持向量机的使用

本节将从 iris 数据集创建一个线性分类器。如前所述,用花萼宽度和花萼长度的特征可以创建一个线性二值分类器来预测是否为山鸢尾花。

4.2.1 开始

为了在 TensorFlow 上实现一个 soft margin 支持向量机,我们将实现特殊的损失函数,公式如下:

$$\frac{1}{n}\sum_{i=1}^{n}\max(0, 1-y_i(Ax_i-b))+a\|A\|^2$$

其中,A 是斜率向量,b 是截距,x_i 是输入向量,y_i 是实际分类(-1 或者 1)。a 是软分类器的正则参数。

4.2.2 动手做

1. 导入必要的编程库,包括导入 scikit learn 的 datasets 库来访问 iris 数据集,代码如下:

```
import matplotlib.pyplot as plt
import numpy as np
import tensorflow as tf
from sklearn import datasets
```

> 安装 scikit learn 可使用:$pip install-U scikit-learn。注意,也可以使用 Anaconda 来安装。

2. 创建一个计算图会话,加载需要的数据集。注意,加载 iris 数据集的第一列和第四列特征变量,其为花萼长度和花萼宽度。加载目标变量时,山鸢尾花为 1,否则为 –1,代码如下:

```
sess = tf.Session()
iris = datasets.load_iris()
x_vals = np.array([[x[0], x[3]] for x in iris.data])
y_vals = np.array([1 if y==0 else -1 for y in iris.target])
```

3. 分割数据集为训练集和测试集。我们将评估训练集和测试集训练的准确度,因为我们知道这个数据集是线性可分的,所以期待在两个数据集上得到 100% 的准确度,代码如下:

```
train_indices = np.random.choice(len(x_vals),
round(len(x_vals)*0.8), replace=False)
test_indices = np.array(list(set(range(len(x_vals))) -
set(train_indices)))
x_vals_train = x_vals[train_indices]
x_vals_test = x_vals[test_indices]
y_vals_train = y_vals[train_indices]
y_vals_test = y_vals[test_indices]
```

4. 设置批量大小、占位符和模型变量。对于这个支持向量机算法，我们希望用非常大的批量大小来帮助其收敛。可以想象一下，非常小的批量大小会使得最大间隔线缓慢跳动。在理想情况下，也应该缓慢减小学习率，但是这已经足够了。A 变量的形状是 2×1，因为有花萼长度和花萼宽度两个变量，代码如下：

```
batch_size = 100

x_data = tf.placeholder(shape=[None, 2], dtype=tf.float32)
y_target = tf.placeholder(shape=[None, 1], dtype=tf.float32)

A = tf.Variable(tf.random_normal(shape=[2,1]))
b = tf.Variable(tf.random_normal(shape=[1,1]))
```

5. 声明模型输出。对于正确分类的数据点，如果数据点是山鸢尾花，则返回的数值大于或者等于 1；否则返回的数值小于或者等于 -1，代码如下：

```
model_output = tf.subtract(tf.matmul(x_data, A), b)
```

6. 声明最大间隔损失函数。首先，我们将声明一个函数来计算向量的 L2 范数。接着增加间隔参数 α。声明分类器损失函数，并把前面两项加在一起，代码如下：

```
l2_norm = tf.reduce_sum(tf.square(A))
alpha = tf.constant([0.1])
classification_term = tf.reduce_mean(tf.maximum(0., tf.subtract(1., tf.multiply(model_output, y_target))))
loss = tf.add(classification_term, tf.multiply(alpha, l2_norm))
```

7. 声明预测函数和准确度函数，用来评估训练集和测试集训练的准确度，代码如下：

```
prediction = tf.sign(model_output)
accuracy = tf.reduce_mean(tf.cast(tf.equal(prediction, y_target), tf.float32))
```

8. 声明优化器函数，并初始化模型变量，代码如下：

```
my_opt = tf.train.GradientDescentOptimizer(0.01)
train_step = my_opt.minimize(loss)

init = tf.global_variables_initializer()
sess.run(init)
```

9. 开始遍历迭代训练模型，记录训练集和测试集训练的损失和准确度，代码如下：

```
loss_vec = []
train_accuracy = []
test_accuracy = []
for i in range(500):
    rand_index = np.random.choice(len(x_vals_train), size=batch_size)
    rand_x = x_vals_train[rand_index]
    rand_y = np.transpose([y_vals_train[rand_index]])
    sess.run(train_step, feed_dict={x_data: rand_x, y_target: rand_y})
    temp_loss = sess.run(loss, feed_dict={x_data: rand_x, y_target: rand_y})
    loss_vec.append(temp_loss)
```

```
        train_acc_temp = sess.run(accuracy, feed_dict={x_data:
x_vals_train, y_target: np.transpose([y_vals_train])})
        train_accuracy.append(train_acc_temp)
        test_acc_temp = sess.run(accuracy, feed_dict={x_data:
x_vals_test, y_target: np.transpose([y_vals_test])})
        test_accuracy.append(test_acc_temp)
        if (i+1)%100==0:
            print('Step #' + str(i+1) + ' A = ' + str(sess.run(A)) + '
b = ' + str(sess.run(b)))
            print('Loss = ' + str(temp_loss))
```

10. 训练过程中前面脚本的输出结果如下：

```
Step #100 A = [[-0.10763293]
 [-0.65735245]] b = [[-0.68752676]]
Loss = [ 0.48756418]
Step #200 A = [[-0.0650763 ]
 [-0.89443302]] b = [[-0.73912662]]
Loss = [ 0.38910741]
Step #300 A = [[-0.02090022]
 [-1.12334013]] b = [[-0.79332656]]
Loss = [ 0.28621092]
Step #400 A = [[ 0.03189624]
 [-1.34912157]] b = [[-0.8507266]]
Loss = [ 0.22397576]
Step #500 A = [[ 0.05958777]
 [-1.55989814]] b = [[-0.9000265]]
Loss = [ 0.20492229]
```

11. 为了绘制输出结果图，需要抽取系数，分割 x_vals 为山鸢尾花（I. setosa）和非山鸢尾花（non-I. setosa），代码如下：

```
[[a1], [a2]] = sess.run(A)
[[b]] = sess.run(b)
slope = -a2/a1
y_intercept = b/a1

x1_vals = [d[1] for d in x_vals]

best_fit = []
for i in x1_vals:
    best_fit.append(slope*i+y_intercept)

setosa_x = [d[1] for i,d in enumerate(x_vals) if y_vals[i]==1]
setosa_y = [d[0] for i,d in enumerate(x_vals) if y_vals[i]==1]
not_setosa_x = [d[1] for i,d in enumerate(x_vals) if y_vals[i]==-1]
not_setosa_y = [d[0] for i,d in enumerate(x_vals) if y_vals[i]==-1]
```

12. 下面是代码绘制数据的线性分类器、准确度和损失图（见图 4-2），代码如下：

```
plt.plot(setosa_x, setosa_y, 'o', label='I. setosa')
plt.plot(not_setosa_x, not_setosa_y, 'x', label='Non-setosa')
plt.plot(x1_vals, best_fit, 'r-', label='Linear Separator',
linewidth=3)
plt.ylim([0, 10])
plt.legend(loc='lower right')
plt.title('Sepal Length vs Petal Width')
plt.xlabel('Petal Width')
```

```
plt.ylabel('Sepal Length')
plt.show()

plt.plot(train_accuracy, 'k-', label='Training Accuracy')
plt.plot(test_accuracy, 'r--', label='Test Accuracy')
plt.title('Train and Test Set Accuracies')
plt.xlabel('Generation')
plt.ylabel('Accuracy')
plt.legend(loc='lower right')
plt.show()

plt.plot(loss_vec, 'k-')
plt.title('Loss per Generation')
plt.xlabel('Generation')
plt.ylabel('Loss')
plt.show()
```

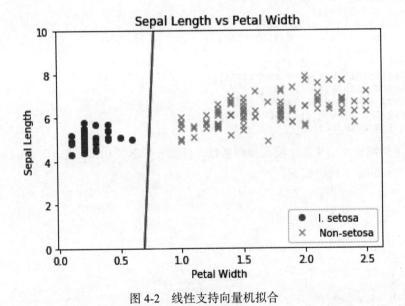

图 4-2 线性支持向量机拟合

> 使用 TensorFlow 实现 SVD 算法可能导致每次运行的结果不尽相同。原因包括训练集和测试集的随机分割，每批训练的批量大小不同，在理想情况下每次迭代后学习率缓慢减小。

从图 4-3 中可以看出，训练集和测试集迭代训练。由于两类目标是线性可分的，我们得到准确度是 100%。迭代 500 次的最大间隔如图 4-4 所示。

4.2.3　工作原理

本节使用最大间隔损失函数实现了线性支持向量机算法模型。

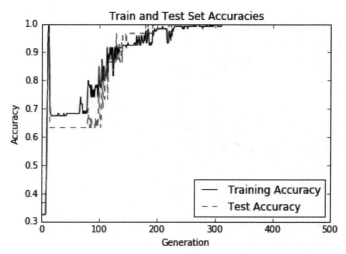

图 4-3　训练集和测试集迭代的准确度。由于两类目标是线性可分的，得到准确度是 100%

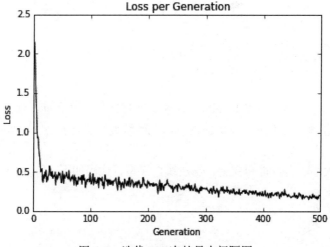

图 4-4　迭代 500 次的最大间隔图

4.3　弱化为线性回归

支持向量机可以用来拟合线性回归。本节将展示如何在 TensorFlow 实现支持向量机用来拟合线性回归。

4.3.1　开始

相同的最大间隔（maximum margin）的概念应用到线性回归拟合。代替最大化分割两类目标是，最大化分割包含大部分的数据点 (x, y)。我们将用相同的 iris 数据集，展示用刚才的概念来进行花萼长度与花瓣宽度之间的线性拟合。

相关的损失函数类似于 max $(0, |y_i - (Ax_i + b)| - \varepsilon)$。这里，$\varepsilon$ 是间隔宽度的一半，这意味着如果一个数据点在该区域，则损失等于0。

4.3.2 动手做

1. 导入必要的编程库，创建一个计算图会话，加载 iris 数据集。然后分割数据集为训练集和测试集，并且可视化相应的损失函数，代码如下：

```
import matplotlib.pyplot as plt
import numpy as np
import tensorflow as tf
from sklearn import datasets
sess = tf.Session()
iris = datasets.load_iris()
x_vals = np.array([x[3] for x in iris.data])
y_vals = np.array([y[0] for y in iris.data])
train_indices = np.random.choice(len(x_vals),
round(len(x_vals)*0.8), replace=False)
test_indices = np.array(list(set(range(len(x_vals))) -
set(train_indices)))
x_vals_train = x_vals[train_indices]
x_vals_test = x_vals[test_indices]
y_vals_train = y_vals[train_indices]
y_vals_test = y_vals[test_indices]
```

对于这个例子，我们分割数据集为训练集和测试集。有时也经常分割为三个数据集，还包括验证集。我们用验证集验证训练过的模型是否过拟合。

2. 声明批量大小、占位符和变量，创建线性模型，代码如下：

```
batch_size = 50

x_data = tf.placeholder(shape=[None, 1], dtype=tf.float32)
y_target = tf.placeholder(shape=[None, 1], dtype=tf.float32)
A = tf.Variable(tf.random_normal(shape=[1,1]))
b = tf.Variable(tf.random_normal(shape=[1,1]))

model_output = tf.add(tf.matmul(x_data, A), b)
```

3. 声明损失函数。该损失函数如前所述，实现时 $\varepsilon = 0.5$。注意，ε 是损失函数的一部分，其允许 soft margin 代替 hard margin，代码如下：

```
epsilon = tf.constant([0.5])
loss = tf.reduce_mean(tf.maximum(0.,
tf.subtract(tf.abs(tf.subtract(model_output, y_target)), epsilon)))
```

4. 创建一个优化器，初始化变量，代码如下：

```
my_opt = tf.train.GradientDescentOptimizer(0.075)
train_step = my_opt.minimize(loss)
```

```
init = tf.global_variables_initializer()
sess.run(init)
```

5. 现在开始200次迭代训练，保存训练集和测试集损失函数，后续用来绘图，代码如下：

```
train_loss = []
test_loss = []
for i in range(200):
    rand_index = np.random.choice(len(x_vals_train), size=batch_size)
    rand_x = np.transpose([x_vals_train[rand_index]])
    rand_y = np.transpose([y_vals_train[rand_index]])
    sess.run(train_step, feed_dict={x_data: rand_x, y_target: rand_y})
    temp_train_loss = sess.run(loss, feed_dict={x_data: np.transpose([x_vals_train]), y_target: np.transpose([y_vals_train])})
    train_loss.append(temp_train_loss)
    temp_test_loss = sess.run(loss, feed_dict={x_data: np.transpose([x_vals_test]), y_target: np.transpose([y_vals_test])})
    test_loss.append(temp_test_loss)
    if (i+1)%50==0:
        print('-----------')
        print('Generation: ' + str(i))
        print('A = ' + str(sess.run(A)) + ' b = ' + str(sess.run(b)))
        print('Train Loss = ' + str(temp_train_loss))
        print('Test Loss = ' + str(temp_test_loss))
```

6. 下面是迭代训练输出结果：

```
Generation: 50
A = [[ 2.20651722]] b = [[ 2.71290684]]
Train Loss = 0.609453
Test Loss = 0.460152
-----------
Generation: 100
A = [[ 1.6440177]] b = [[ 3.75240564]]
Train Loss = 0.242519
Test Loss = 0.208901
-----------
Generation: 150
A = [[ 1.27711761]] b = [[ 4.3149066]]
Train Loss = 0.108192
Test Loss = 0.119284
-----------
Generation: 200
A = [[ 1.05271816]] b = [[ 4.53690529]]
Train Loss = 0.0799957
Test Loss = 0.107551
```

7. 现在抽取系数，获取最佳拟合直线的截距。为了后续绘图，这里也获取间隔宽度值，代码如下：

```
[[slope]] = sess.run(A)
[[y_intercept]] = sess.run(b)
[width] = sess.run(epsilon)
```

```
best_fit = []
best_fit_upper = []
best_fit_lower = []
for i in x_vals:
  best_fit.append(slope*i+y_intercept)
  best_fit_upper.append(slope*i+y_intercept+width)
  best_fit_lower.append(slope*i+y_intercept-width)
```

8. 最后，绘制数据点和拟合直线，以及训练集和测试集损失，代码如下（对应的图见图 4-5）：

```
plt.plot(x_vals, y_vals, 'o', label='Data Points')
plt.plot(x_vals, best_fit, 'r-', label='SVM Regression Line', linewidth=3)
plt.plot(x_vals, best_fit_upper, 'r--', linewidth=2)
plt.plot(x_vals, best_fit_lower, 'r--', linewidth=2)
plt.ylim([0, 10])
plt.legend(loc='lower right')
plt.title('Sepal Length vs Petal Width')
plt.xlabel('Petal Width')
plt.ylabel('Sepal Length')
plt.show()
plt.plot(train_loss, 'k-', label='Train Set Loss')
plt.plot(test_loss, 'r--', label='Test Set Loss')
plt.title('L2 Loss per Generation')
plt.xlabel('Generation')
plt.ylabel('L2 Loss')
plt.legend(loc='upper right')
plt.show()
```

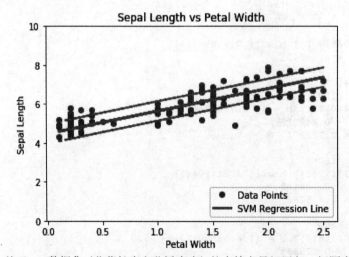

图 4-5 基于 iris 数据集（花萼长度和花瓣宽度）的支持向量机回归，间隔宽度为 0.5

图 4-6 是模型训练迭代中训练集和测试集的训练损失：

4.3.3 工作原理

直观地讲，我们认为 SVM 回归算法试图把更多的数据点拟合到直线两边 2ε 宽度的间隔内。这时拟合的直线对于 ε 参数更有意义。如果选择太小的 ε 值，SVM 回归算法在间隔宽度内不能拟合更多的数据点；如果选择太大的 ε 值，将有许多条直线能够在间隔宽度内拟合所有的数据点。作者更倾向于选取更小的 ε 值，因为在间隔宽度附近的数据点比远处的数据点贡献更少的损失。

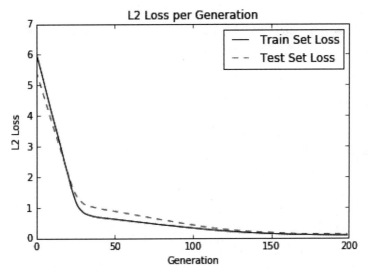

图 4-6　每次迭代的支持向量机回归的损失值（训练集和测试集）

4.4　TensorFlow 上核函数的使用

上节介绍的是用 SVM 算法线性分割数据集。如果想分割非线性数据集，该如何改变线性分类器映射到数据集？答案是，改变 SVM 损失函数中的核函数。本节将详细阐述如何调整核函数，并且分割非线性可分的数据集。

4.4.1　开始

我们将介绍支持向量机算法中核函数的使用。在 4.2 节中，采用特殊损失函数来解决 soft margin 问题。另外一种解决 soft margin 的方法是对偶优化问题，线性支持向量机问题的对偶表达式为：

$$\max \sum_{i=1}^{n} b_i - \frac{1}{2} \sum_{i=1}^{n} \sum_{j=1}^{n} y_i b_i (x_i \cdot x_j) y_j b_j$$

其中：

$$\sum_{i=1}^{n} b_i y_i = 0 \text{ 和 } 0 \leqslant b_i \leqslant \frac{1}{2ny}$$

上述表达式中，模型变量是向量 b。在理想情况下，b 向量是稀疏向量，iris 数据集相关的支持向量仅仅取 1 和 -1 附近的值。数据点向量以 x_i 表示，目标值（1 或者 -1）以 y_i 表示。

在前述方程里的核函数是点积，$x_i \cdot x_j$，其为线性核函数。该核函数是以数据点（i, j）的点积填充的方阵。

替代数据点间的点积，可以将其扩展到更复杂的函数更高维度。这看似不怎么复杂，但是如果选择函数 k，其需满足如下条件：

$$k(x_i, x_j) = \varphi(x_i) \cdot \varphi(x_j)$$

这里，k 称为核函数。最广为人知的核函数之一是高斯核函数（也称为，径向基核函数或者 RBF 核函数），该核函数用下面的方程描述：

$$k(x_i, x_j) = e^{-y} \|x_i - x_j\|^2$$

为了用该核函数预测，假设观测数据点 p_i，代入上述核函数等式中：

$$k(x_i, p_j) = e^{-y} \|x_i - p_j\|^2$$

在本节中，我们将讨论如何实现高斯核函数。注意，这里将用合适的线性核函数实现来替代。为了显示高斯核函数比线性核函数更合适，使用的数据集是程序生成的模拟数据。

4.4.2 动手做

1. 导入必要的编程库，创建一个计算图会话，代码如下：

```
import matplotlib.pyplot as plt
import numpy as np
import tensorflow as tf
from sklearn import datasets
sess = tf.Session()
```

2. 生成模拟数据。生成的数据是两个同心圆数据，每个不同的环代表不同的类，确保只有类 –1 或者 1。为了让绘图方便，这里将每类数据分成 x 值和 y 值，代码如下：

```
(x_vals, y_vals) = datasets.make_circles(n_samples=500, factor=.5,
noise=.1)
y_vals = np.array([1 if y==1 else -1 for y in y_vals])
class1_x = [x[0] for i,x in enumerate(x_vals) if y_vals[i]==1]
class1_y = [x[1] for i,x in enumerate(x_vals) if y_vals[i]==1]
class2_x = [x[0] for i,x in enumerate(x_vals) if y_vals[i]==-1]
class2_y = [x[1] for i,x in enumerate(x_vals) if y_vals[i]==-1]
```

3. 声明批量大小、占位符，创建模型变量 b。对于 SVM 算法，为了让每次迭代训练不波动，得到一个稳定的训练模型，这时批量大小得取值更大。注意，本例为预测数据点声明有额外的占位符。最后创建彩色的网格来可视化不同的区域代表不同的类别，代码如下：

```
batch_size = 250
x_data = tf.placeholder(shape=[None, 2], dtype=tf.float32)
y_target = tf.placeholder(shape=[None, 1], dtype=tf.float32)
prediction_grid = tf.placeholder(shape=[None, 2], dtype=tf.float32)
b = tf.Variable(tf.random_normal(shape=[1,batch_size]))
```

4. 创建高斯核函数。该核函数用矩阵操作来表示，代码如下：

```
gamma = tf.constant(-50.0)
dist = tf.reduce_sum(tf.square(x_data), 1)
```

```
dist = tf.reshape(dist, [-1,1])
sq_dists = tf.add(tf.subtract(dist, tf.multiply(2.,
tf.matmul(x_data, tf.transpose(x_data)))), tf.transpose(dist))
my_kernel = tf.exp(tf.multiply(gamma, tf.abs(sq_dists)))
```

注意，在 sq_dists 中应用广播加法和减法操作。

线性核函数可以表示为：my_kernel = tf.matmul(x_data, tf.transpose(x_data))。

5. 声明在本节一开始提到的对偶问题。为了最大化，这里采用最小化损失函数的负数：tf.neg()，代码如下：

```
model_output = tf.matmul(b, my_kernel)
first_term = tf.reduce_sum(b)
b_vec_cross = tf.matmul(tf.transpose(b), b)
y_target_cross = tf.matmul(y_target, tf.transpose(y_target))
second_term = tf.reduce_sum(tf.multiply(my_kernel,
tf.multiply(b_vec_cross, y_target_cross)))
loss = tf.negative(tf.subtract(first_term, second_term))
```

6. 创建预测函数和准确度函数。先创建一个预测核函数，类似于步骤 4，但用预测数据点的核函数代替步骤 4 中用模拟数据点的核函数。预测值是模型输出的符号函数值，代码如下：

```
rA = tf.reshape(tf.reduce_sum(tf.square(x_data), 1),[-1,1])
rB = tf.reshape(tf.reduce_sum(tf.square(prediction_grid),
1),[-1,1])
pred_sq_dist = tf.add(tf.subtract(rA, tf.multiply(2.,
tf.matmul(x_data, tf.transpose(prediction_grid)))),
tf.transpose(rB))
pred_kernel = tf.exp(tf.multiply(gamma, tf.abs(pred_sq_dist)))

prediction_output =
tf.matmul(tf.multiply(tf.transpose(y_target),b), pred_kernel)
prediction = tf.sign(prediction_output-
tf.reduce_mean(prediction_output))
accuracy = tf.reduce_mean(tf.cast(tf.equal(tf.squeeze(prediction),
tf.squeeze(y_target)), tf.float32))
```

为了实现线性预测核函数，将预测核函数改为：pred_kernel = tf.matmul(x_data, tf.transpose(prediction_grid))。

7. 创建优化器函数，初始化所有的变量，代码如下：

```
my_opt = tf.train.GradientDescentOptimizer(0.001)
train_step = my_opt.minimize(loss)
init = tf.global_variables_initializer()
sess.run(init)
```

8. 开始迭代训练。这里会记录每次迭代的损失向量和批量训练的准确度。当计算准确度时，需要为三个占位符赋值，其中，x_data 数据会被赋值两次来得到数据点的预测值，代码如下：

```
loss_vec = []
batch_accuracy = []
for i in range(500):
    rand_index = np.random.choice(len(x_vals), size=batch_size)
    rand_x = x_vals[rand_index]
    rand_y = np.transpose([y_vals[rand_index]])
    sess.run(train_step, feed_dict={x_data: rand_x, y_target: rand_y})
    temp_loss = sess.run(loss, feed_dict={x_data: rand_x, y_target: rand_y})
    loss_vec.append(temp_loss)
    acc_temp = sess.run(accuracy, feed_dict={x_data: rand_x,
                                             y_target: rand_y,
prediction_grid:rand_x})
    batch_accuracy.append(acc_temp)
    if (i+1)%100==0:
        print('Step #' + str(i+1))
        print('Loss = ' + str(temp_loss))
```

9. 输出结果如下：

```
Step #100
Loss = -28.0772
Step #200
Loss = -3.3628
Step #300
Loss = -58.862
Step #400
Loss = -75.1121
Step #500
Loss = -84.8905
```

10. 为了能够在整个数据空间可视化分类返回结果，我们将创建预测数据点的网格，并在其上进行预测，代码如下：

```
x_min, x_max = x_vals[:, 0].min() - 1, x_vals[:, 0].max() + 1
y_min, y_max = x_vals[:, 1].min() - 1, x_vals[:, 1].max() + 1
xx, yy = np.meshgrid(np.arange(x_min, x_max, 0.02),
                     np.arange(y_min, y_max, 0.02))
grid_points = np.c_[xx.ravel(), yy.ravel()]
[grid_predictions] = sess.run(prediction, feed_dict={x_data: x_vals,
                                                     y_target: np.transpose([y_vals]),
                                                     prediction_grid: grid_points})
grid_predictions = grid_predictions.reshape(xx.shape)
```

11. 下面绘制预测结果、批量准确度和损失函数：

```
plt.contourf(xx, yy, grid_predictions, cmap=plt.cm.Paired, alpha=0.8)
plt.plot(class1_x, class1_y, 'ro', label='Class 1')
plt.plot(class2_x, class2_y, 'kx', label='Class -1')
plt.legend(loc='lower right')
plt.ylim([-1.5, 1.5])
plt.xlim([-1.5, 1.5])
plt.show()
```

```
plt.plot(batch_accuracy, 'k-', label='Accuracy')
plt.title('Batch Accuracy')
plt.xlabel('Generation')
plt.ylabel('Accuracy')
plt.legend(loc='lower right')
plt.show()

plt.plot(loss_vec, 'k-')
plt.title('Loss per Generation')
plt.xlabel('Generation')
plt.ylabel('Loss')
plt.show()
```

限于篇幅这里只显示训练结果图（见图 4-7 和图 4-8），不过也可以分开运行绘图代码展示误差和准确度。

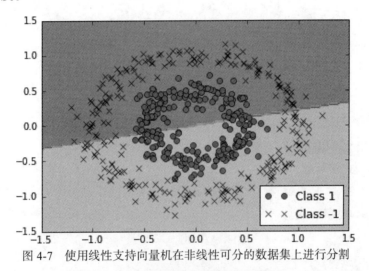

图 4-7　使用线性支持向量机在非线性可分的数据集上进行分割

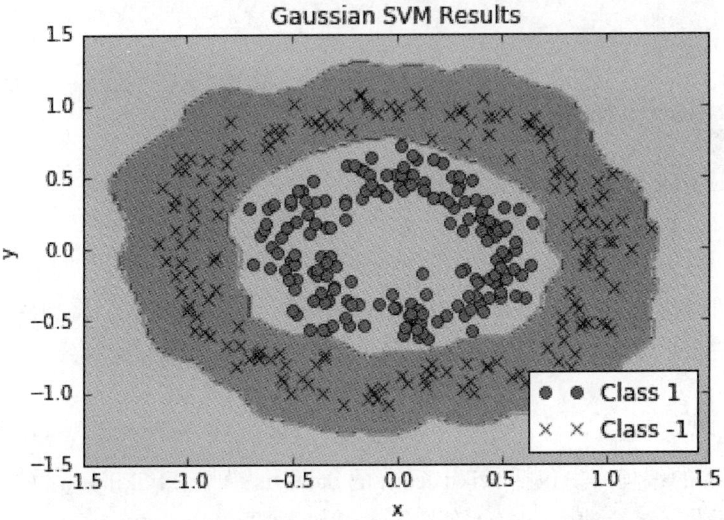

图 4-8　使用非线性的高斯核函数 SVM 在非线性可分的数据集上进行分割

图 4-7 展示的是在非线性可分的数据集上进行线性 SVM 训练。

图 4-8 展示的是非线性可分的数据集上进行非线性的高斯核函数 SVM 训练。

4.4.3 工作原理

上述的代码里有两个重要的部分：如何为 SVM 对偶优化问题完成核函数和损失函数。我们已经实现了线性核函数和高斯核函数，其中高斯核函数能够分割非线性数据集。

我们也应该注意到高斯核函数中有一个参数——gamma。该参数控制数据集分割的弯曲部分的影响程度，一般情况下选择较小值，但是也严重依赖于数据集。在理想情况下，gamma 值是通过统计技术（比如，交叉验证）来确定的。

对于一个新点的预测或者评估，可以使用以下命令 sess.run(prediciton,feed_dict:{x_data:x_vals,y_data:np.transpose([y_vals])})。该命令必须包含原始数据集 x_vals 和 y_vals，原因在于 SVM 由支持向量定义，依据原始数据集明确新点在边缘或者不在边缘上。

4.4.4 延伸学习

有许多核函数可以实现，下面是常用的非线性核函数列表：

☐ 齐次多项式核函数：

$$k(x_i, x_j) = (x_i \cdot x_j)^d$$

☐ 非齐次多项式核函数：

$$k(x_i, x_j) = (x_i \cdot x_j + 1)^d$$

☐ 双曲正切核函数：

$$k(x_i, x_j) = \tanh(ax_i \cdot x_j + k)$$

4.5 用 TensorFlow 实现非线性支持向量机

本节将会应用非线性核函数来分割数据集。

4.5.1 开始

本节使用前面实现的高斯核函数 SVM 来分割真实数据集。我们将加载 iris 数据集，创建一个山鸢尾花（I. setosa）的分类器。后面将会看到各种 gamma 值对分类器的影响。

4.5.2 动手做

1. 导入必要的编程库，包括 scikit learn 的 datasets 模块。scikit learn 的 datasets 模块可以加载 iris 数据集。然后建立一个计算图会话，代码如下：

```
import matplotlib.pyplot as plt
import numpy as np
import tensorflow as tf
from sklearn import datasets
sess = tf.Session()
```

2. 加载 iris 数据集,抽取花萼长度和花瓣宽度,分割每类的 x_vals 值和 y_vals 值,代码如下:

```
iris = datasets.load_iris()
x_vals = np.array([[x[0], x[3]] for x in iris.data])
y_vals = np.array([1 if y==0 else -1 for y in iris.target])
class1_x = [x[0] for i,x in enumerate(x_vals) if y_vals[i]==1]
class1_y = [x[1] for i,x in enumerate(x_vals) if y_vals[i]==1]
class2_x = [x[0] for i,x in enumerate(x_vals) if y_vals[i]==-1]
class2_y = [x[1] for i,x in enumerate(x_vals) if y_vals[i]==-1]
```

3. 声明批量大小(偏向于更大批量大小)、占位符和模型变量 b,代码如下:

```
batch_size = 100

x_data = tf.placeholder(shape=[None, 2], dtype=tf.float32)
y_target = tf.placeholder(shape=[None, 1], dtype=tf.float32)
prediction_grid = tf.placeholder(shape=[None, 2], dtype=tf.float32)

b = tf.Variable(tf.random_normal(shape=[1,batch_size]))
```

4. 声明高斯核函数。该核函数依赖 gamma 值,在本节末尾将展示不同的 gamma 值对分类器的影响,代码如下:

```
gamma = tf.constant(-10.0)
dist = tf.reduce_sum(tf.square(x_data), 1)
dist = tf.reshape(dist, [-1,1])
sq_dists = tf.add(tf.subtract(dist, tf.multiply(2.,
tf.matmul(x_data, tf.transpose(x_data)))), tf.transpose(dist))
my_kernel = tf.exp(tf.multiply(gamma, tf.abs(sq_dists)))
# We now compute the loss for the dual optimization problem, as
follows:
model_output = tf.matmul(b, my_kernel)
first_term = tf.reduce_sum(b)
b_vec_cross = tf.matmul(tf.transpose(b), b)
y_target_cross = tf.matmul(y_target, tf.transpose(y_target))
second_term = tf.reduce_sum(tf.multiply(my_kernel,
tf.multiply(b_vec_cross, y_target_cross)))
loss = tf.negative(tf.subtract(first_term, second_term))
```

5. 为了使用 SVM 进行预测,创建一个预测核函数。然后声明一个准确度函数,其为正确分类的数据点的百分比,代码如下:

```
rA = tf.reshape(tf.reduce_sum(tf.square(x_data), 1),[-1,1])
rB = tf.reshape(tf.reduce_sum(tf.square(prediction_grid),
1),[-1,1])
pred_sq_dist = tf.add(tf.subtract(rA, tf.mul(2., tf.matmul(x_data,
tf.transpose(prediction_grid)))), tf.transpose(rB))
pred_kernel = tf.exp(tf.multiply(gamma, tf.abs(pred_sq_dist)))

prediction_output =
```

```
tf.matmul(tf.multiply(tf.transpose(y_target),b), pred_kernel)
prediction = tf.sign(prediction_output-
tf.reduce_mean(prediction_output))
accuracy = tf.reduce_mean(tf.cast(tf.equal(tf.squeeze(prediction),
tf.squeeze(y_target)), tf.float32))
```

6. 声明优化器函数,初始化变量,代码如下:

```
my_opt = tf.train.GradientDescentOptimizer(0.01)
train_step = my_opt.minimize(loss)
init = tf.initialize_all_variables()
sess.run(init)
```

7. 现在开始迭代训练。迭代 300 次,并保存损失值和批量准确度,代码如下:

```
loss_vec = []
batch_accuracy = []
for i in range(300):
    rand_index = np.random.choice(len(x_vals), size=batch_size)
    rand_x = x_vals[rand_index]
    rand_y = np.transpose([y_vals[rand_index]])
    sess.run(train_step, feed_dict={x_data: rand_x, y_target:
rand_y})
    temp_loss = sess.run(loss, feed_dict={x_data: rand_x, y_target:
rand_y})
    loss_vec.append(temp_loss)
    acc_temp = sess.run(accuracy, feed_dict={x_data: rand_x,
                                              y_target: rand_y,
prediction_grid:rand_x})
    batch_accuracy.append(acc_temp)
```

8. 为了绘制决策边界(Decision Boundary),我们创建一个数据点 (x, y) 的网格,评估预测函数,代码如下:

```
x_min, x_max = x_vals[:, 0].min() - 1, x_vals[:, 0].max() + 1
y_min, y_max = x_vals[:, 1].min() - 1, x_vals[:, 1].max() + 1
xx, yy = np.meshgrid(np.arange(x_min, x_max, 0.02),
                      np.arange(y_min, y_max, 0.02))
grid_points = np.c_[xx.ravel(), yy.ravel()]
[grid_predictions] = sess.run(prediction, feed_dict={x_data:
x_vals,
                                                       y_target:
np.transpose([y_vals]),
                                                      prediction_grid:
grid_points})
grid_predictions = grid_predictions.reshape(xx.shape)
```

9. 限于篇幅,这里仅仅显示如何绘制决策边界。对于 gamma 值的影响和绘图将在下一节介绍,代码如下:

```
plt.contourf(xx, yy, grid_predictions, cmap=plt.cm.Paired,
alpha=0.8)
plt.plot(class1_x, class1_y, 'ro', label='I. setosa')
plt.plot(class2_x, class2_y, 'kx', label='Non-setosa')
plt.title('Gaussian SVM Results on Iris Data')
plt.xlabel('Petal Length')
plt.ylabel('Sepal Width')
plt.legend(loc='lower right')
```

```
plt.ylim([-0.5, 3.0])
plt.xlim([3.5, 8.5])
plt.show()
```

4.5.3 工作原理

这里是四种不同的 gamma 值（1，10，25，100），山鸢尾花（I. setosa）的分类器结果图，如图 4-9 所示。注意，gamma 值越大，每个数据点对分类边界的影响就越大。

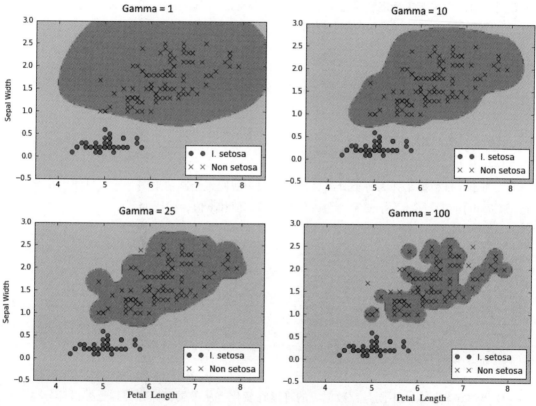

图 4-9 不同 gamma 值的山鸢尾花（I. setosa）的分类器结果图，采用高斯核函数的 SVM

4.6 用 TensorFlow 实现多类支持向量机

我们也能用 SVM 分类多类目标。在本节中，将详细展示一个多类支持向量机分类器训练 iris 数据集来分类三种花。

4.6.1 开始

SVM 算法最初是为二值分类问题设计的，但是也可以通过一些策略使得其能进行多类分类。主要的两种策略是：一对多（one versus all）方法；一对一（one versus one）方法。

一对一方法是在任意两类样本之间设计创建一个二值分类器，然后得票最多的类别即为该未知样本的预测类别。但是当类别（k 类）很多的时候，就必须创建 $k!/(k-2)!2!$ 个分类器，计算的代价还是相当大的。

另外一种实现多类分类器的方法是一对多，其为每类创建一个分类器。最后的预测类别是具有最大 SVM 间隔的类别。本小节将实现该方法。

我们将加载 iris 数据集，使用高斯核函数的非线性多类 SVM 模型。iris 数据集含有三个类别，山鸢尾、变色鸢尾和维吉尼亚鸢尾（I. setosa、I. virginica 和 I. versicolor），我们将为它们创建三个高斯核函数 SVM 来预测。

4.6.2 动手做

1. 导入必要的编程库，创建一个计算图，代码如下：

```
import matplotlib.pyplot as plt
import numpy as np
import tensorflow as tf
from sklearn import datasets
sess = tf.Session()
```

2. 加载 iris 数据集并为每类分离目标值。因为我们想绘制结果图，所以只使用花萼长度和花瓣宽度两个特征。为了便于绘图，也会分离 x 值和 y 值，代码如下：

```
iris = datasets.load_iris()
x_vals = np.array([[x[0], x[3]] for x in iris.data])
y_vals1 = np.array([1 if y==0 else -1 for y in iris.target])
y_vals2 = np.array([1 if y==1 else -1 for y in iris.target])
y_vals3 = np.array([1 if y==2 else -1 for y in iris.target])
y_vals = np.array([y_vals1, y_vals2, y_vals3])
class1_x = [x[0] for i,x in enumerate(x_vals) if iris.target[i]==0]
class1_y = [x[1] for i,x in enumerate(x_vals) if iris.target[i]==0]
class2_x = [x[0] for i,x in enumerate(x_vals) if iris.target[i]==1]
class2_y = [x[1] for i,x in enumerate(x_vals) if iris.target[i]==1]
class3_x = [x[0] for i,x in enumerate(x_vals) if iris.target[i]==2]
class3_y = [x[1] for i,x in enumerate(x_vals) if iris.target[i]==2]
```

3. 与 4.5 节最大的不同是，数据集的维度在变化，从单类目标分类到三类目标分类。我们将利用矩阵传播和 reshape 技术一次性计算所有的三类 SVM。注意，由于一次性计算所有分类，y_target 占位符的维度是 [3, None]，模型变量 b 初始化大小为 [3, batch_size]，代码如下：

```
batch_size = 50

x_data = tf.placeholder(shape=[None, 2], dtype=tf.float32)
y_target = tf.placeholder(shape=[3, None], dtype=tf.float32)
prediction_grid = tf.placeholder(shape=[None, 2], dtype=tf.float32)

b = tf.Variable(tf.random_normal(shape=[3,batch_size]))
```

4. 计算高斯核函数。因为该核函数只依赖 x_data，所以代码与 4.5 节没有区别，代码如下：

```
gamma = tf.constant(-10.0)
dist = tf.reduce_sum(tf.square(x_data), 1)
dist = tf.reshape(dist, [-1,1])
sq_dists = tf.add(tf.subtract(dist, tf.multiply(2.,
tf.matmul(x_data, tf.transpose(x_data)))), tf.transpose(dist))
my_kernel = tf.exp(tf.multiply(gamma, tf.abs(sq_dists)))
```

5. 最大的变化是批量矩阵乘法。最终的结果是三维矩阵，并且需要传播矩阵乘法。所以数据矩阵和目标矩阵需要预处理，比如 $x^T \cdot x$ 操作需额外增加一个维度。这里创建一个函数来扩展矩阵维度，然后进行矩阵转置，接着调用 TensorFlow 的 tf.batch_matmul() 函数，代码如下：

```
def reshape_matmul(mat):
    v1 = tf.expand_dims(mat, 1)
    v2 = tf.reshape(v1, [3, batch_size, 1])
    return tf.batch_matmul(v2, v1)
```

6. 计算对偶损失函数，代码如下：

```
model_output = tf.matmul(b, my_kernel)
first_term = tf.reduce_sum(b)
b_vec_cross = tf.matmul(tf.transpose(b), b)
y_target_cross = reshape_matmul(y_target)

second_term = tf.reduce_sum(tf.multiply(my_kernel,
tf.multiply(b_vec_cross, y_target_cross)),[1,2])
loss = tf.reduce_sum(tf.negative(tf.subtract(first_term,
second_term)))
```

7. 现在创建预测核函数。要注意 reduce_sum() 函数，这里我们并不想聚合三个 SVM 预测，所以需要通过第二个参数告诉 TensorFlow 求和哪几个，代码如下：

```
rA = tf.reshape(tf.reduce_sum(tf.square(x_data), 1),[-1,1])
rB = tf.reshape(tf.reduce_sum(tf.square(prediction_grid),
1),[-1,1])
pred_sq_dist = tf.add(tf.subtract(rA, tf.multiply(2.,
tf.matmul(x_data, tf.transpose(prediction_grid)))),
tf.transpose(rB))
pred_kernel = tf.exp(tf.multiply(gamma, tf.abs(pred_sq_dist)))
```

8. 实现预测核函数后，我们创建预测函数。与 4.5 节不同的是，不再对模型输出进行 sign() 运算。因为这里实现的是一对多方法，所以预测值是分类器有最大返回值的类别。使用 TensorFlow 的内建函数 argmax() 来实现该功能，代码如下：

```
prediction_output = tf.matmul(tf.mul(y_target,b), pred_kernel)
prediction = tf.arg_max(prediction_output-
tf.expand_dims(tf.reduce_mean(prediction_output,1), 1), 0)
accuracy = tf.reduce_mean(tf.cast(tf.equal(prediction,
tf.argmax(y_target,0)), tf.float32))
```

9. 准备好核函数、损失函数和预测函数后，该声明优化器函数和初始化变量了，代码如下：

```
my_opt = tf.train.GradientDescentOptimizer(0.01)
train_step = my_opt.minimize(loss)
```

```
init = tf.global_variables_initializer()
sess.run(init)
```

10. 该算法收敛得相当快，所以迭代训练的次数不要超过 100 次，代码如下：

```
loss_vec = []
batch_accuracy = []
for i in range(100):
    rand_index = np.random.choice(len(x_vals), size=batch_size)
    rand_x = x_vals[rand_index]
    rand_y = y_vals[:,rand_index]
    sess.run(train_step, feed_dict={x_data: rand_x, y_target: rand_y})
    temp_loss = sess.run(loss, feed_dict={x_data: rand_x, y_target: rand_y})
    loss_vec.append(temp_loss)
    acc_temp = sess.run(accuracy, feed_dict={x_data: rand_x, y_target: rand_y, prediction_grid:rand_x})
    batch_accuracy.append(acc_temp)
    if (i+1)%25==0:
        print('Step #' + str(i+1))
        print('Loss = ' + str(temp_loss))

Step #25
Loss = -2.8951
Step #50
Loss = -27.9612
Step #75
Loss = -26.896
Step #100
Loss = -30.2325
```

11. 创建数据点的预测网格，运行预测函数，代码如下：

```
x_min, x_max = x_vals[:, 0].min() - 1, x_vals[:, 0].max() + 1
y_min, y_max = x_vals[:, 1].min() - 1, x_vals[:, 1].max() + 1
xx, yy = np.meshgrid(np.arange(x_min, x_max, 0.02),
                     np.arange(y_min, y_max, 0.02))
grid_points = np.c_[xx.ravel(), yy.ravel()]
grid_predictions = sess.run(prediction, feed_dict={x_data: rand_x,
                                                   y_target: rand_y,
                                                   prediction_grid: grid_points})
grid_predictions = grid_predictions.reshape(xx.shape)
```

12. 绘制训练结果、批量准确度和损失函数（见图 4-10）。为了简便，只显示训练结果：

```
plt.contourf(xx, yy, grid_predictions, cmap=plt.cm.Paired, alpha=0.8)
plt.plot(class1_x, class1_y, 'ro', label='I. setosa')
plt.plot(class2_x, class2_y, 'kx', label='I. versicolor')
plt.plot(class3_x, class3_y, 'gv', label='I. virginica')
plt.title('Gaussian SVM Results on Iris Data')
plt.xlabel('Petal Length')
plt.ylabel('Sepal Width')
plt.legend(loc='lower right')
plt.ylim([-0.5, 3.0])
plt.xlim([3.5, 8.5])
```

```
plt.show()

plt.plot(batch_accuracy, 'k-', label='Accuracy')
plt.title('Batch Accuracy')
plt.xlabel('Generation')
plt.ylabel('Accuracy')
plt.legend(loc='lower right')
plt.show()

plt.plot(loss_vec, 'k-')
plt.title('Loss per Generation')
plt.xlabel('Generation')
plt.ylabel('Loss')
plt.show()
```

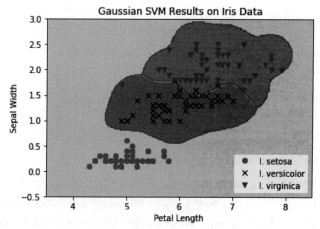

图 4-10　山鸢尾花（I.Setosa）非线性高斯 SVM 模型的多分类（三类）结果，其中 gamma 值为 10

4.6.3　工作原理

　　本节的重点是改变 SVM 算法一次性优化三类 SVM 模型。模型参数 b 通过增加一个维度来计算三个模型。我们可以看到，使用 TensorFlow 内建功能可以轻松扩展算法到多类的相似算法。

　　第 5 章将涉及最近领域法，它是对于预测问题非常有效的一种方法。

CHAPTER 5

第 5 章

最近邻域法

本章主要关注最近邻域法，以及在 TensorFlow 中如何实现该算法。首先介绍最近邻域法和不同形式的实现方法，然后在本章结尾处将最近邻域法应用于地址匹配和图像识别中。学完本章将掌握以下知识点：
- 最近邻域法的使用
- 如何度量文本距离
- 用 TensorFlow 实现混合距离
- 用 TensorFlow 实现地址匹配
- 用 TensorFlow 实现图像识别

注意，本章的代码可以在 GitHub（https://github.com/nfmcclure/tensorflow_cookbook）和 Packt 代码库（https://github.com/PacktPublishing/TensorFlow-Machine-Learning-Cookbook-Second-Edition）上访问。

5.1 简介

最近邻域算法的思想很简单，其先将训练集看作训练模型，然后基于新数据点与训练集的距离来预测新数据点。最直观的最近邻域算法是让预测值与最接近的训练数据集作为同一类。但是大部分样本数据集包含一定程度的噪声，更通用的方法是 k 个邻域的加权平均，该方法称为 k 最近邻域法（k-nearest neighbor，k-NN）。

假设样本训练集 $(x_1, x_2, \cdots, x_n)$，对应的目标值 $(y_1, y_2, \cdots, y_n)$，通过最近邻域法预测数据点 z。预测的实际方法取决于我们是想做回归训练（连续型 y_i）还是分类训练（离散型 y_i）。

对于离散型分类目标，预测值由到预测数据点的加权距离的最大投票方案决定，公式如下：

$$f(z) = \max_j \sum_{i=1}^{k} \varphi(d_{ij}) I_{ij}$$

其中，预测函数 $f(z)$ 是所有分类 j 上的最大加权值。预测数据点到训练数据点 i 的加权

距离用 $\varphi(d_{ij})$ 表示。I_{ij} 是指示函数，表示数据点 i 是否属于分类 j，当数据点 i 属于分类 j 时，指示函数为 1，否则为 0，同时 k 是距离预测数据点最近的训练数据个数。

对于连续回归训练目标，预测值是所有 k 个最近邻域数据点到预测数据点的加权平均，公式如下：

$$f(z) = \frac{1}{k}\sum_{i=1}^{k} \varphi(d_i)$$

明显地，预测值严重依赖距离度量（d）方式的选择。常用的距离度量是 L1 范数和 L2 范数。公式如下：

$$d_{L1}(x_i, x_j) = |x_i - x_j| = |x_{i1} - x_{j1}| + |x_{i2} - x_{j2}| + \cdots$$

$$d_{L2}(x_i, x_j) = \|x_i - x_j\| = \sqrt{(x_{i1} - x_{j1})^2 + (x_{i2} - x_{j2})^2 + \cdots}$$

距离度量方式可选择性广，但是在本节，将使用 L1 范数和 L2 范数，也会使用编辑距离和文本距离。

我们也需要选择如何加权距离。最直观的方式是用距离本身来加权，即加权权重为 1。考虑到更近的数据点对预测数据点的预测值影响应该更小，因而最通用的加权方式是距离的归一化倒数。下一节将实现该方法。

注意，k-NN 算法是一种聚合的方法。对于回归算法来说，需要计算邻域的加权平均距离，因而预测值将比实际目标值的特征更平缓。影响的程度将取决于 k 值，该值是算法中的邻域的个数。

5.2 最近邻域法的使用

本小节开始介绍最近邻域法的实现，并应用到房价的预测。也许这是学习最近邻域法最好的方式，因为我们将处理数值化的特征和连续型目标。

5.2.1 开始

为了展示在 TensorFlow 中如何运用最近邻域法预测，我们将进行波士顿房价数据集训练。这里将利用几个特征的函数来预测平均邻域房价。

我们将从训练好的模型的训练数据集中找到预测数据点的最近邻域，并对实际值进行加权平均。

5.2.2 动手做

1. 导入必要的编程库，创建一个计算图会话。我们将使用 Python 的 requests 模块，从 UCI 机器学习仓库加载所需的波士顿房价数据集，代码如下：

```python
import matplotlib.pyplot as plt
import numpy as np
import tensorflow as tf
import requests

sess = tf.Session()
```

2. 使用 requests 模块加载数据集，代码如下：

```python
housing_url =
'https://archive.ics.uci.edu/ml/machine-learning-databases/housing/
housing.data'
housing_header = ['CRIM', 'ZN', 'INDUS', 'CHAS', 'NOX', 'RM',
'AGE', 'DIS', 'RAD', 'TAX', 'PTRATIO', 'B', 'LSTAT', 'MEDV']
cols_used = ['CRIM', 'INDUS', 'NOX', 'RM', 'AGE', 'DIS', 'TAX',
'PTRATIO', 'B', 'LSTAT']
num_features = len(cols_used)
# Request data
housing_file = requests.get(housing_url)
# Parse Data
housing_data = [[float(x) for x in y.split(' ') if len(x)>=1] for y
in housing_file.text.split('n') if len(y)>=1]
```

3. 分离数据集为特征依赖的数据集和特征无关的数据集。我们将预测最后一个变量——MEDV，该值为一组房价中的平均值。由于非相关特征或者二值特征，在本例中不使用 ZN、CHAS 和 RAD 这几个特征，代码如下：

```python
y_vals = np.transpose([np.array([y[13] for y in housing_data])])
x_vals = np.array([[x for i,x in enumerate(y) if housing_header[i]
in cols_used] for y in housing_data])

x_vals = (x_vals - x_vals.min(0)) / x_vals.ptp(0)
```

4. 分离 x_vals 值和 y_vals 值为训练数据集和测试数据集。随机选择 80% 的行作为训练集，剩下的 20% 数据行作为测试集，代码如下：

```python
train_indices = np.random.choice(len(x_vals),
round(len(x_vals)*0.8), replace=False)
test_indices = np.array(list(set(range(len(x_vals))) -
set(train_indices)))
x_vals_train = x_vals[train_indices]
x_vals_test = x_vals[test_indices]
y_vals_train = y_vals[train_indices]
y_vals_test = y_vals[test_indices]
```

5. 声明 k 值和批量大小：

```python
k = 4
batch_size=len(x_vals_test)
```

6. 声明占位符。注意，本例中没有训练模型变量，算法模型完全是通过训练集决定的，代码如下：

```python
x_data_train = tf.placeholder(shape=[None, num_features],
dtype=tf.float32)
x_data_test = tf.placeholder(shape=[None, num_features],
dtype=tf.float32)
```

```
y_target_train = tf.placeholder(shape=[None, 1], dtype=tf.float32)
y_target_test = tf.placeholder(shape=[None, 1], dtype=tf.float32)
```

7. 为批量测试集创建距离函数,这里使用 L1 范数距离,代码如下:

```
distance = tf.reduce_sum(tf.abs(tf.subtract(x_data_train,
tf.expand_dims(x_data_test,1))), reduction_indices=2)
```

 注意,L2 范数距离函数也经常使用,代码为:
```
distance = tf.sqrt(tf.reduce_sum(tf.square(tf.
sub(x_data_train, tf.expand_dims(x_data_test,1))),
reduction_indices=1))
```

8. 创建预测函数。在本例中,将使用 top_k() 函数,其以张量的方式返回最大值的值和索引。因为需要找到最小距离的索引,所以将对最大距离取负。声明预测函数和目标值的均方误差(MSE),代码如下:

```
top_k_xvals, top_k_indices = tf.nn.top_k(tf.negative(distance),
k=k)
x_sums = tf.expand_dims(tf.reduce_sum(top_k_xvals, 1),1)
x_sums_repeated = tf.matmul(x_sums,tf.ones([1, k], tf.float32))
x_val_weights =
tf.expand_dims(tf.divide(top_k_xvals,x_sums_repeated), 1)

top_k_yvals = tf.gather(y_target_train, top_k_indices)
prediction = tf.squeeze(tf.batch_matmul(x_val_weights,top_k_yvals),
squeeze_dims=[1])
mse = tf.divide(tf.reduce_sum(tf.square(tf.subtract(prediction,
y_target_test))), batch_size)
```

9. 进行测试,对测试数据进行循环检验,存储预测值和准确度,代码如下:

```
num_loops = int(np.ceil(len(x_vals_test)/batch_size))

for i in range(num_loops):
    min_index = i*batch_size
    max_index = min((i+1)*batch_size,len(x_vals_train))
    x_batch = x_vals_test[min_index:max_index]
    y_batch = y_vals_test[min_index:max_index]
    predictions = sess.run(prediction, feed_dict={x_data_train:
x_vals_train, x_data_test: x_batch, y_target_train: y_vals_train,
y_target_test: y_batch})
    batch_mse = sess.run(mse, feed_dict={x_data_train:
x_vals_train, x_data_test: x_batch, y_target_train: y_vals_train,
y_target_test: y_batch})

    print('Batch #' + str(i+1) + ' MSE: ' +
str(np.round(batch_mse,3)))
```

Batch #1 MSE: 23.153

10. 下面通过直方图(见图 5-1)来比较实际值和预测值。因为本例使用的是平均方法,所以在预测目标值最大和最小极值时遇到问题,代码如下:

```
bins = np.linspace(5, 50, 45)
plt.hist(predictions, bins, alpha=0.5, label='Prediction')
plt.hist(y_batch, bins, alpha=0.5, label='Actual')
plt.title('Histogram of Predicted and Actual Values')
plt.xlabel('Med Home Value in $1,000s')
plt.ylabel('Frequency')
plt.legend(loc='upper right')
plt.show()
```

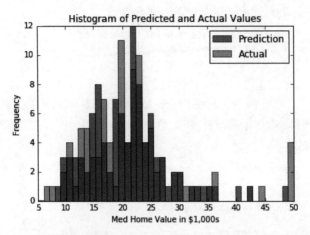

图 5-1　预测值和实际值对比的直方图（k-NN 算法，其中 $k=4$）

在算法中 k 的取值是个难点。对于图 5-1 中的预测，我们选用 $k=4$ 进行模型训练。选取该值的原因是，通过交叉验证，其使得 MSE 最低。如果使用交叉验证来选取多个 k 值对比，我们将看到 $k=4$ 时，MSE 值最小，见图 5-2。同时，也值得绘制预测值随 k 值变化的预测方差，因为其可以显示平均更多的邻域数据点时预测方差也会减小。

5.2.3　工作原理

在最近邻域算法中，模型是训练数据集，所以在训练模型中没有任何变量训练。只有一个参数 k，通过交叉验证法最小化 MSE 来确定。

5.2.4　延伸学习

对于 k-NN 算法的权重，我们选择的是距离的本身。另外也有一些其他的权重选取方法，常用平方距离的倒数作为权重。

5.3　如何度量文本距离

除了处理数值外，最近邻域法还广泛应用于其他领域。只要有方法来度量距离，即可应用最近邻域算法。本节将介绍如何使用 TensorFlow 度量文本距离。

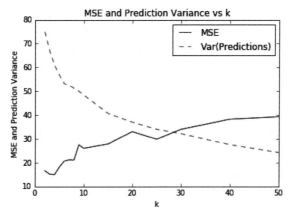

图 5-2 随 k 值变化的 k-NN 预测的 MSE 值。我们也绘制测试集预测值的预测方差。注意，随着 k 值增大，预测方差变小

5.3.1 开始

本节将展示如何使用 TensorFlow 的文本距离度量——字符串间的编辑距离（Levenshtein 距离）。这将对后文扩展最近邻域法很有帮助。

Levenshtein 距离是指由一个字符串转换成另一个字符串所需的最少编辑操作次数。允许的编辑操作包括插入一个字符、删除一个字符和将一个字符替换成另一个字符。本节将使用 TensorFlow 的内建函数 edit_distance() 求解 Levenshtein 距离。本节将展示该函数的使用，后续章节会有应用。

 注意，TensorFlow 的内建函数 edit_distance() 仅仅接受稀疏张量。因此，我们得把字符串转换成稀疏张量。

5.3.2 动手做

1. 加载 TensorFlow，初始化一个计算图会话，代码如下：

```
import tensorflow as tf
sess = tf.Session()
```

2. 展示如何计算两个单词 'bear' 和 'beer' 间的编辑距离。用 Python 的 list() 函数创建字符列表，然后将列表映射为一个三维稀疏矩阵。TensorFlow 的 tf.SparseTensor() 函数需指定字符索引、矩阵形状和张量中的非零值。编辑距离计算时，指定 normalize=False 表示计算总的编辑距离；指定 normalize=True 表示计算归一化编辑距离（通过编辑距离除以第二个单词的长度进行归一化），代码如下：

```
hypothesis = list('bear')
truth = list('beers')
h1 = tf.SparseTensor([[0,0,0], [0,0,1], [0,0,2], [0,0,3]],
```

```
                        hypothesis, [1,1,1])
t1 = tf.SparseTensor([[0,0,0], [0,0,1], [0,0,1], [0,0,3],[0,0,4]],
truth, [1,1,1])

print(sess.run(tf.edit_distance(h1, t1, normalize=False)))
```

`[[ 2.]]`

 TensorFlow 文档把两个字符串处理为参考字符串（hypothesis）和真实字符串（ground truth）。本例标记为 h 张量和 t 张量。

TensorFlow 的 SparseTensorValue() 函数是创建稀疏张量的方法，要传入所需创建的稀疏张量的索引、值和形状。

3. 下面演示比较两个单词 bear 和 beer 与另一个单词 beers。为了做比较，需要重复 beers 使得比较的单词有相同的数量，代码如下：

```
hypothesis2 = list('bearbeer')
truth2 = list('beersbeers')
h2 = tf.SparseTensor([[0,0,0], [0,0,1], [0,0,2], [0,0,3], [0,1,0],
[0,1,1], [0,1,2], [0,1,3]], hypothesis2, [1,2,4])
t2 = tf.SparseTensor([[0,0,0], [0,0,1], [0,0,2], [0,0,3], [0,0,4],
[0,1,0], [0,1,1], [0,1,2], [0,1,3], [0,1,4]], truth2, [1,2,5])

print(sess.run(tf.edit_distance(h2, t2, normalize=True)))
```

`[[ 0.40000001  0.2       ]]`

4. 下面介绍一个例子，讲解另外一种更有效地比较一个单词集合与单个单词的方法。事先为参考字符串和真实字符串创建索引和字符列表，代码如下：

```
hypothesis_words = ['bear','bar','tensor','flow']
truth_word = ['beers']
num_h_words = len(hypothesis_words)
h_indices = [[xi, 0, yi] for xi,x in enumerate(hypothesis_words)
for yi,y in enumerate(x)]
h_chars = list(''.join(hypothesis_words))
h3 = tf.SparseTensor(h_indices, h_chars, [num_h_words,1,1])
truth_word_vec = truth_word*num_h_words
t_indices = [[xi, 0, yi] for xi,x in enumerate(truth_word_vec) for
yi,y in enumerate(x)]
t_chars = list(''.join(truth_word_vec))
t3 = tf.SparseTensor(t_indices, t_chars, [num_h_words,1,1])

print(sess.run(tf.edit_distance(h3, t3, normalize=True)))
```

```
[[ 0.40000001]
 [ 0.60000002]
 [ 0.80000001]
 [ 1.        ]]
```

5. 下一步将展示如何用占位符来计算两个单词列表间的编辑距离。基本思路是一样的，不同的是现在用 SparseTensorValue() 替代先前的稀疏张量。首先，创建一个函数，该函数根据单词列表输出稀疏张量，代码如下：

```
def create_sparse_vec(word_list):
    num_words = len(word_list)
    indices = [[xi, 0, yi] for xi,x in enumerate(word_list) for yi,y in enumerate(x)]
    chars = list(''.join(word_list))
    return(tf.SparseTensorValue(indices, chars, [num_words,1,1]))

hyp_string_sparse = create_sparse_vec(hypothesis_words)
truth_string_sparse = create_sparse_vec(truth_word*len(hypothesis_words))

hyp_input = tf.sparse_placeholder(dtype=tf.string)
truth_input = tf.sparse_placeholder(dtype=tf.string)

edit_distances = tf.edit_distance(hyp_input, truth_input, normalize=True)

feed_dict = {hyp_input: hyp_string_sparse,
             truth_input: truth_string_sparse}
print(sess.run(edit_distances, feed_dict=feed_dict))

[[ 0.40000001]
 [ 0.60000002]
 [ 0.80000001]
 [ 1.        ]]
```

5.3.3　工作原理

本节展示了使用 TensorFlow 计算文本距离的几种方法。这对文本特征数据进行最近邻域法训练非常有帮助。本章后续的地址匹配应用中也会用到这些方法。

5.3.4　延伸学习

这里讨论一下文本距离的度量方式，下面是各种文本距离的定义，假设字符串 s1 和 s2。

名　　称	描　　述	公　　式
汉明距离（Hamming distance）	两个等长字符串中对应位置的不同字符的个数	$D(s_1, s_2) = \Sigma_i I_i$，其中 I 是等长字符串的指示函数
余弦距离（Cosine distance）	不同 k-gram 的点积除以不同 k-gram 的 L2 范数	$D(s_1, s_2) = 1 - \dfrac{k(s_1) \cdot k(s_2)}{\|k(s_1)\| \|k(s_2)\|}$
Jaccard 距离（Jaccard distance）	两个字符串中相同字符数除以所有字符数	$D(s_1, s_2) = \dfrac{|s_1 \cap s_2|}{|s_1 \cup s_2|}$

5.4　用 TensorFlow 实现混合距离计算

当处理的数据观测点有多种特征时，我们应该意识到不同的特征应该用不同的归一化方式来缩放。本节将用此思路来优化房价预测值。

5.4.1 开始

扩展最近邻域法进行多维度缩放。在本例中，我们将展示如何扩展多变量的距离函数。特别地，我们将扩展距离函数为特征变量的函数。

加权距离函数的关键是使用加权权重矩阵。包含矩阵操作的距离函数的表达式如下：

$$D(x, y) = \sqrt{(x-y)^T A (x-y)}$$

其中，A 是对角权重矩阵，用来对每个特征的距离度量进行调整。

在本节中，我们将试着优化波士顿房价数据集的 MSE。该数据集的特征维度不同，所以缩放后的距离函数对最近邻域法有利。

5.4.2 动手做

1.导入必要的编程库，创建一个计算图会话，代码如下：

```
import matplotlib.pyplot as plt
import numpy as np
import tensorflow as tf
import requests
sess = tf.Session()
```

2.加载数据集，存储为 numpy 数组。再次提醒，我们只使用某些列来预测，不使用 id 变量或者方差非常小的变量，代码如下：

```
housing_url =
'https://archive.ics.uci.edu/ml/machine-learning-databases/housing/
housing.data'
housing_header = ['CRIM', 'ZN', 'INDUS', 'CHAS', 'NOX', 'RM',
'AGE', 'DIS', 'RAD', 'TAX', 'PTRATIO', 'B', 'LSTAT', 'MEDV']
cols_used = ['CRIM', 'INDUS', 'NOX', 'RM', 'AGE', 'DIS', 'TAX',
'PTRATIO', 'B', 'LSTAT']
num_features = len(cols_used)
housing_file = requests.get(housing_url)
housing_data = [[float(x) for x in y.split(' ') if len(x)>=1] for y
in housing_file.text.split('\n') if len(y)>=1]
y_vals = np.transpose([np.array([y[13] for y in housing_data])])
x_vals = np.array([[x for i,x in enumerate(y) if housing_header[i]
in cols_used] for y in housing_data])
```

3.用 min-max 缩放法缩放 x 值到 0 和 1 之间，代码如下：

```
x_vals = (x_vals - x_vals.min(0)) / x_vals.ptp(0)
```

4.创建对角权重矩阵，该矩阵提供归一化的距离度量，其值为特征的标准差，代码如下：

```
weight_diagonal = x_vals.std(0)
weight_matrix = tf.cast(tf.diag(weight_diagonal), dtype=tf.float32)
```

5.分割数据集为训练集和测试集。声明 k 值，该值为最近邻域的数量。设置批量大小为测试集大小，代码如下：

```
train_indices = np.random.choice(len(x_vals),
round(len(x_vals)*0.8), replace=False)
test_indices = np.array(list(set(range(len(x_vals))) -
set(train_indices)))
x_vals_train = x_vals[train_indices]
x_vals_test = x_vals[test_indices]
y_vals_train = y_vals[train_indices]
y_vals_test = y_vals[test_indices]
k = 4
batch_size=len(x_vals_test)
```

6. 声明所需的占位符。占位符有四个，分别是训练集和测试集的 x 值输入和 y 目标输入，代码如下：

```
x_data_train = tf.placeholder(shape=[None, num_features],
dtype=tf.float32)
x_data_test = tf.placeholder(shape=[None, num_features],
dtype=tf.float32)
y_target_train = tf.placeholder(shape=[None, 1], dtype=tf.float32)
y_target_test = tf.placeholder(shape=[None, 1], dtype=tf.float32)
```

7. 声明距离函数。为了使可读性更好，我们将距离函数分解。注意，本例需要 tf.tile 函数为权重矩阵指定 batch_size 维度扩展，使用 batch_matmul() 函数进行批量矩阵乘法，代码如下：

```
subtraction_term = tf.subtract(x_data_train,
tf.expand_dims(x_data_test,1))
first_product = tf.batch_matmul(subtraction_term,
tf.tile(tf.expand_dims(weight_matrix,0), [batch_size,1,1]))
second_product = tf.batch_matmul(first_product,
tf.transpose(subtraction_term, perm=[0,2,1]))
distance = tf.sqrt(tf.batch_matrix_diag_part(second_product))
```

8. 计算完每个测试数据点的距离，需要返回 k-NN 法的前 k 个最近邻域（使用 tf.nn.top_k() 函数）。因为 tf.nn.top_k() 函数返回最大值，而我们需要的是最小距离，所以转换成返回距离负值的最大值。然后将前 k 个最近邻域的距离进行加权平均做预测，代码如下：

```
top_k_xvals, top_k_indices = tf.nn.top_k(tf.neg(distance), k=k)
x_sums = tf.expand_dims(tf.reduce_sum(top_k_xvals, 1),1)
x_sums_repeated = tf.matmul(x_sums,tf.ones([1, k], tf.float32))
x_val_weights = tf.expand_dims(tf.div(top_k_xvals,x_sums_repeated),
1)
top_k_yvals = tf.gather(y_target_train, top_k_indices)
prediction = tf.squeeze(tf.batch_matmul(x_val_weights,top_k_yvals),
squeeze_dims=[1])
```

9. 计算预测值的 MSE，评估训练模型，代码如下：

```
mse = tf.divide(tf.reduce_sum(tf.square(tf.subtract(prediction,
y_target_test))), batch_size)
```

10. 遍历迭代训练批量测试数据，每次迭代计算其 MSE，代码如下：

```
num_loops = int(np.ceil(len(x_vals_test)/batch_size))
for i in range(num_loops):
    min_index = i*batch_size
    max_index = min((i+1)*batch_size,len(x_vals_train))
    x_batch = x_vals_test[min_index:max_index]
```

```
        y_batch = y_vals_test[min_index:max_index]
        predictions = sess.run(prediction, feed_dict={x_data_train:
x_vals_train, x_data_test: x_batch, y_target_train: y_vals_train,
y_target_test: y_batch})
        batch_mse = sess.run(mse, feed_dict={x_data_train:
x_vals_train, x_data_test: x_batch, y_target_train: y_vals_train,
y_target_test: y_batch})
        print('Batch #' + str(i+1) + ' MSE: ' +
str(np.round(batch_mse,3)))

Batch #1 MSE: 21.322
```

11. 为了最终对比，我们绘制测试数据集的房价分布和测试集上的预测值的分布（见图5-3），代码如下：

```
bins = np.linspace(5, 50, 45)
plt.hist(predictions, bins, alpha=0.5, label='Prediction')
plt.hist(y_batch, bins, alpha=0.5, label='Actual')
plt.title('Histogram of Predicted and Actual Values')
plt.xlabel('Med Home Value in $1,000s')
plt.ylabel('Frequency')
plt.legend(loc='upper right')
plt.show()
```

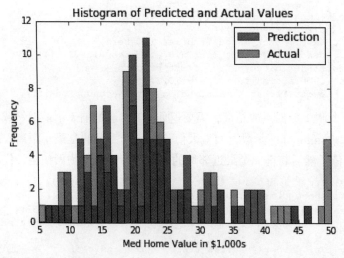

图 5-3　波士顿房价的预测值和实际值的分布直方图。本图中对特征进行了距离缩放

5.4.3　工作原理

通过对特征的距离进行缩放，减小测试数据集的MSE。本例中使用的是特征的标准差因子来缩放距离函数，并且对top k 邻域进行加权平均作为距离函数来进行房价预测。

5.4.4　延伸学习

缩放因子可以加强或者减弱最近邻域距离计算中特征的权重，这更符合实际特征的作用。

5.5 用 TensorFlow 实现地址匹配

完成数值距离和文本距离后，现在我们将花点时间结合两者来度量既包含文本特征又包含数值特征的数据观测点间的距离。

5.5.1 开始

最近邻域算法应用在地址匹配上是非常有效的。地址匹配是一种记录匹配，其匹配的地址涉及多个数据集。在地址匹配中，地址中有许多打印错误，不同的城市或者不同的邮政编码，但是指向同一个地址。使用最近邻域算法综合地址信息的数值部分和字符部分可以帮助鉴定实际相同的地址。

本例将生成两个模拟数据集，每个数据集包含街道地址和邮政编码。其中，有一个数据集的街道地址有大量的打印错误。我们将准确的地址数据集作为"金标准"，为每个有打印错误的地址返回一个最接近的地址，采用综合字符距离（街道）和数值距离（邮政编码）的距离函数度量地址间的相似程度。

代码的第一部分是生成模拟数据集。第二部分是训练测试数据集，从训练数据集中返回最接近的地址。

5.5.2 动手做

1. 先导入必要的编程库，代码如下：

```
import random
import string
import numpy as np
import tensorflow as tf
```

2. 创建参考数据集。为了显示简洁的输出，每个数据集仅仅由 10 个地址组成，不过更多数据量也适用，代码如下：

```
n = 10 street_names = ['abbey', 'baker', 'canal', 'donner', 'elm']
street_types = ['rd', 'st', 'ln', 'pass', 'ave']
rand_zips = [random.randint(65000,65999) for i in range(5)]
numbers = [random.randint(1, 9999) for i in range(n)]
streets = [random.choice(street_names) for i in range(n)]
street_suffs = [random.choice(street_types) for i in range(n)]
zips = [random.choice(rand_zips) for i in range(n)]
full_streets = [str(x) + ' ' + y + ' ' + z for x,y,z in
zip(numbers, streets, street_suffs)]
reference_data = [list(x) for x in zip(full_streets,zips)]
```

3. 为了创建一个测试数据集，我们需要一个随机创建"打印错误"的字符串函数，然后返回结果字符串，代码如下：

```
def create_typo(s, prob=0.75):
    if random.uniform(0,1) < prob:
        rand_ind = random.choice(range(len(s)))
        s_list = list(s)
```

```
            s_list[rand_ind]=random.choice(string.ascii_lowercase)
            s = ''.join(s_list)
    return s

typo_streets = [create_typo(x) for x in streets]
typo_full_streets = [str(x) + ' ' + y + ' ' + z for x,y,z in
zip(numbers, typo_streets, street_suffs)]
test_data = [list(x) for x in zip(typo_full_streets,zips)]
```

4. 初始化一个计算图会话，声明所需的占位符。本例需要四个占位符，每个测试集和参考集需一个地址和邮政编码占位符，代码如下：

```
sess = tf.Session()
test_address = tf.sparse_placeholder( dtype=tf.string)
test_zip = tf.placeholder(shape=[None, 1], dtype=tf.float32)
ref_address = tf.sparse_placeholder(dtype=tf.string)
ref_zip = tf.placeholder(shape=[None, n], dtype=tf.float32)
```

5. 声明数值的邮政编码距离和地址字符串的编辑距离，代码如下：

```
zip_dist = tf.square(tf.subtract(ref_zip, test_zip))
address_dist = tf.edit_distance(test_address, ref_address,
normalize=True)
```

6. 把邮政编码距离和地址距离转换成相似度。当两个输入完全一致时该相似度为1；当它们完全不一致时为0。对于邮政编码相似度，其计算方式为：最大邮政编码减去该邮政编码，然后除以邮政编码范围（即最大邮政编码减去最小邮政编码的差值）。对于地址相似度，其值已经是0到1之间的值，所以直接用1减去其编辑距离大小，代码如下：

```
zip_max = tf.gather(tf.squeeze(zip_dist), tf.argmax(zip_dist, 1))
zip_min = tf.gather(tf.squeeze(zip_dist), tf.argmin(zip_dist, 1))
zip_sim = tf.divide(tf.subtract(zip_max, zip_dist),
tf.subtract(zip_max, zip_min))
address_sim = tf.subtract(1., address_dist)
```

7. 结合上面两个相似度函数，并对其进行加权平均。在本例中，地址和邮政编码的权重设为相等（即，各为0.5），但我们也可以根据每个特征的信誉度来调整权重，然后返回参考集最大相似度的索引，代码如下：

```
address_weight = 0.5
zip_weight = 1. - address_weight
weighted_sim = tf.add(tf.transpose(tf.multiply(address_weight,
address_sim)), tf.multiply(zip_weight, zip_sim))
top_match_index = tf.argmax(weighted_sim, 1)
```

8. 为了在TensorFlow中使用编辑距离，我们必须把地址字符串转换成稀疏向量。在本章前面的小节中，我们创建过下面的函数来进行稀疏矩阵的转换，代码如下：

```
def sparse_from_word_vec(word_vec):
    num_words = len(word_vec)
    indices = [[xi, 0, yi] for xi,x in enumerate(word_vec) for yi,y
in enumerate(x)]
    chars = list(''.join(word_vec))
    # Now we return our sparse vector
    return tf.SparseTensorValue(indices, chars, [num_words,1,1])
```

9. 分离参考集中的地址和邮政编码，然后在遍历迭代训练中为占位符赋值，代码如下：

```
reference_addresses = [x[0] for x in reference_data]
reference_zips = np.array([[x[1] for x in reference_data]])
```

10. 利用步骤 8 中创建的函数将参考地址转换为稀疏矩阵，代码如下：

```
sparse_ref_set = sparse_from_word_vec(reference_addresses)
```

11. 遍历循环测试集的每项，返回参考集中最接近项的索引，打印出测试集和参考集的每项。正如下面所看到的，模拟数据集训练的结果不错，代码如下：

```
for i in range(n):
    test_address_entry = test_data[i][0]
    test_zip_entry = [[test_data[i][1]]]
    # Create sparse address vectors
    test_address_repeated = [test_address_entry] * n
    sparse_test_set = sparse_from_word_vec(test_address_repeated)
    feeddict={test_address: sparse_test_set,
              test_zip: test_zip_entry,
              ref_address: sparse_ref_set,
              ref_zip: reference_zips}
    best_match = sess.run(top_match_index, feed_dict=feeddict)
    best_street = reference_addresses[best_match[0]]
    [best_zip] = reference_zips[0][best_match]
    [[test_zip_]] = test_zip_entry
    print('Address: ' + str(test_address_entry) + ', ' + str(test_zip_))
    print('Match  : ' + str(best_street) + ', ' + str(best_zip))
```

输出结果如下：

```
Address: 8659 beker ln, 65463
Match  : 8659 baker ln, 65463
Address: 1048 eanal ln, 65681
Match  : 1048 canal ln, 65681
Address: 1756 vaker st, 65983
Match  : 1756 baker st, 65983
Address: 900 abbjy pass, 65983
Match  : 900 abbey pass, 65983
Address: 5025 canal rd, 65463
Match  : 5025 canal rd, 65463
Address: 6814 elh st, 65154
Match  : 6814 elm st, 65154
Address: 3057 cagal ave, 65463
Match  : 3057 canal ave, 65463
Address: 7776 iaker ln, 65681
Match  : 7776 baker ln, 65681
Address: 5167 caker rd, 65154
Match  : 5167 baker rd, 65154
Address: 8765 donnor st, 65154
Match  : 8765 donner st, 65154
```

5.5.3 工作原理

解决地址匹配问题时会遇到很多困难，比如，权重大小和如何归一化距离，这些都得根据实际数据选择解决方法。可能处理地址的方法与刚才的方法不同。比如，把街道地址细化成省市、城市、街道地址和街道号码。

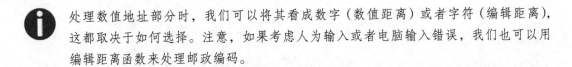

处理数值地址部分时，我们可以将其看成数字（数值距离）或者字符（编辑距离），这都取决于如何选择。注意，如果考虑人为输入或者电脑输入错误，我们也可以用编辑距离函数来处理邮政编码。

为了感受"打印错误"对结果的影响，我们鼓励读者去调整"打印错误"的字符串函数来增加"打印错误"的数量或者错误出现的频率，增加数据样本集的大小来查看算法的效果。

5.6　用 TensorFlow 实现图像识别

最近邻域算法也常用于图像识别。图像识别领域的"Hello World"数据集是 MNIST 手写数字样本数据集。后续的章节会使用该数据集进行各种神经网络图像识别算法训练，这也和非神经网络算法的结果进行对比。

5.6.1　开始

MNIST 手写数字样本数据集由上万张 28×28 像素、已标注的图片组成。虽然该数据集不大，但是其包含有 784 个特征可供最近邻域算法训练。我们将计算这类分类问题的最近邻域预测，选用最近 k 邻域（本例中，$k=4$）模型。

5.6.2　动手做

1. 导入必要的编程库。注意，导入 PIL（Python Image Library）模块绘制预测输出结果。TensorFlow 中有内建的函数加载 MNIST 手写数字样本数据集，代码如下：

```
import random
import numpy as np
import tensorflow as tf
import matplotlib.pyplot as plt
from PIL import Image
from tensorflow.examples.tutorials.mnist import input_data
```

2. 创建一个计算图会话，加载 MNIST 手写数字数据集，并指定 one-hot 编码，代码如下：

```
sess = tf.Session()
mnist = input_data.read_data_sets("MNIST_data/", one_hot=True)
```

one-hot 编码是分类类别的数值化，这样更有利于后续的数值计算。本例包含 10 个类别（数字 0 到 9），采用长度为 10 的 0-1 向量表示。例如，类别"0"表示为向量：1,0,0,0,0,0,0,0,0,0，类别"1"表示为向量：0,1,0,0,0,0,0,0,0,0，等等。

3. 由于 MNIST 手写数字数据集较大，直接计算成千上万个输入的 784 个特征之间的距离是比较困难的，所以本例会抽样成小数据集进行训练。对测试集也进行抽样处理，为了后续绘图方便，选择测试集数量可以被 6 整除。我们将绘制最后批次的 6 张图片来查看效果，代码如下：

```
train_size = 1000
test_size = 102

rand_train_indices = np.random.choice(len(mnist.train.images),
train_size, replace=False)
rand_test_indices = np.random.choice(len(mnist.test.images),
test_size, replace=False)
x_vals_train = mnist.train.images[rand_train_indices]
x_vals_test = mnist.test.images[rand_test_indices]
y_vals_train = mnist.train.labels[rand_train_indices]
y_vals_test = mnist.test.labels[rand_test_indices]
```

4. 声明 k 值和批量大小：

```
k = 4
batch_size=6
```

5. 现在在计算图中开始初始化占位符，并赋值，代码如下：

```
x_data_train = tf.placeholder(shape=[None, 784], dtype=tf.float32)
x_data_test = tf.placeholder(shape=[None, 784], dtype=tf.float32)
y_target_train = tf.placeholder(shape=[None, 10], dtype=tf.float32)
y_target_test = tf.placeholder(shape=[None, 10], dtype=tf.float32)
```

6. 声明距离度量函数。本例使用 L1 范数（即绝对值）作为距离函数，代码如下：

```
distance = tf.reduce_sum(tf.abs(tf.subtract(x_data_train,
tf.expand_dims(x_data_test,1))), reduction_indices=2)
```

> 注意，我们也可以把距离函数定义为 L2 范数。对应的代码为：distance = tf.sqrt(tf.reduce_sum(tf.square(tf.sub(x_data_train, tf.expand_dims(x_data_test, 1))), reduction_indices = 1))。

7. 找到最接近的 top k 图片和预测模型。在数据集的 one-hot 编码索引上进行预测模型计算，然后统计发生的数量，最终预测为数量最多的类别，代码如下：

```
top_k_xvals, top_k_indices = tf.nn.top_k(tf.negative(distance),
k=k)
prediction_indices = tf.gather(y_target_train, top_k_indices)
count_of_predictions = tf.reduce_sum(prediction_indices,
reduction_indices=1)
prediction = tf.argmax(count_of_predictions)
```

8. 在测试集上遍历迭代运行，计算预测值，并将结果存储，代码如下：

```
num_loops = int(np.ceil(len(x_vals_test)/batch_size))
test_output = []
actual_vals = []
```

```
for i in range(num_loops):
    min_index = i*batch_size
    max_index = min((i+1)*batch_size,len(x_vals_train))
    x_batch = x_vals_test[min_index:max_index]
    y_batch = y_vals_test[min_index:max_index]
    predictions = sess.run(prediction, feed_dict={x_data_train:
x_vals_train, x_data_test: x_batch, y_target_train: y_vals_train,
y_target_test: y_batch})
    test_output.extend(predictions)
    actual_vals.extend(np.argmax(y_batch, axis=1))
```

9. 现在已经保存了实际值和预测返回值,下面计算模型训练准确度。不过该结果会因为测试数据集和训练数据集的随机抽样而变化,但是其准确度约为 80%～90%,代码如下:

```
accuracy = sum([1./test_size for i in range(test_size) if
test_output[i]==actual_vals[i]])
print('Accuracy on test set: ' + str(accuracy))
Accuracy on test set: 0.8333333333333325
```

10. 绘制最后批次的计算结果(见图 5-4),代码如下:

```
actuals = np.argmax(y_batch, axis=1)
Nrows = 2
Ncols = 3
for i in range(len(actuals)):
    plt.subplot(Nrows, Ncols, i+1)
    plt.imshow(np.reshape(x_batch[i], [28,28]), cmap='Greys_r')
    plt.title('Actual: ' + str(actuals[i]) + ' Pred: ' +
str(predictions[i]), fontsize=10)
    frame = plt.gca()
    frame.axes.get_xaxis().set_visible(False)
    frame.axes.get_yaxis().set_visible(False)
```

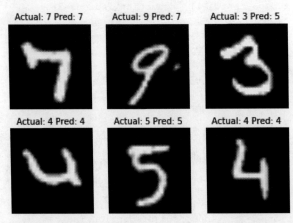

图 5-4　最近邻域算法预测的最后批次的六张图片。我们能看到,并不是每张图片都获得到正确的结果

5.6.3　工作原理

假设有足够的计算时间和计算资源,我们能让训练数据集和测试数据集足够大。该方

法可以增加准确度，也是预防过拟合的最普通的方法之一。并且最近邻域算法说明了理想的 k 值需进一步选择，k 值的选择一般通过在数据集上进行交叉验证获得。

5.6.4 延伸学习

我们也能使用最近邻域算法评估用户书写的数字，具体见代码仓库（https://github.com/nfmcclure/tensorflow_cookbook）。

本章讨论了如何使用 kNN 算法进行回归训练和分类。也列举了各种距离函数的使用方法，以及如何混合使用各种函数。我们鼓励读者开发更多不同的距离度量函数、权重和 k 值来提高算法的准确度。

CHAPTER 6
第 6 章

神经网络算法

在本章中，我们将介绍神经网络算法及其在 TensorFlow 中的实现。后续大部分章节会基于神经网络算法，所以学习如何在 TensorFlow 中实现神经网络算法非常重要。我们将从介绍神经网络的基本概念讲起，之后介绍多层神经网络算法，在最后一节，将创建一个神经网络来学习井字棋。

通过本章的学习，你将掌握以下知识点：
- 实现门函数
- 使用门函数和激励函数
- TensorFlow 实现单层神经网络
- TensorFlow 实现神经网络常见层
- 使用多层神经网络
- 线性预测模型的优化
- TensorFlow 基于神经网络实现井字棋

读者可以在代码仓库（https://github.com/nfmcclure/tensorflow_cookbook）和 Packt 代码库（https://github.com/PacktPublishing/TensorFlow-Machine-Learning-Cookbook-Second-Edition）获取本章的所有代码。

6.1 简介

神经网络算法在识别图像和语音、识别手写、理解文本、图像分割、对话系统、自动驾驶等领域不断打破纪录。其中，部分应用将在后续章节进行介绍。在此之前，本章的重点在于神经网络这一简单易用的机器学习算法。

神经网络算法的概念已出现几十年，但是它仅仅在最近由于计算能力（计算处理、算法效率和数据集大小）的提升能训练大规模网络才获得新的发展。

神经网络算法是对输入数据矩阵进行一系列的基本操作。这些操作通常包括非线性函数的加法和乘法，在 3.9 节中有所使用。逻辑回归算法是斜率与特征点积求和后进行非线性

sigmoid 函数计算。神经网络算法表达形式更通用，允许任意形式的基本操作和非线性函数的结合，包括绝对值、最大值、最小值等。

神经网络算法的一个重要的"黑科技"是"反向传播"。反向传播是一种基于学习率和损失函数返回值来更新模型变量的过程。我们在第 3 章和第 4 章中使用反向传播方法来更新模型变量。

神经网络算法另外一个重要的特性是非线性激励函数。因为大部分神经网络算法仅仅是加法操作和乘法操作的结合，所以它们不能进行非线性数据样本集的模型训练。为了解决该问题，我们在神经网络算法中使用非线性激励函数，这将使得神经网络算法能够解决大部分非线性的问题。

记住，如前面见过的大部分算法，神经网络算法对所选择的超参数是敏感的。在本章中，我们将看到不同的学习率、损失函数和优化过程对模型训练的影响。

这里有一些神经网络算法的深入学习资料。

关于反向传播的文章"Efficient BackProp"，Yann LeCun 等著，PDF 版地址：http://yann.lecun.com/exdb/publis/pdf/lecun-98b.pdf。

斯坦福大学课程 CS231，《Convolutional Neural Networks for Visual Recognition》，课件地址：http://cs231n.stanford.edu/。

斯坦福大学课程 CS224d，《Deep Learning for Natural Language Processing》，课件地址：http://cs224d.stanford.edu/。

MIT 出版的《Deep Learning》，Goodfellow 等著，地址：http://www.deeplearningbook.org。

在线书籍《Neural Networks and Deep Learning》，Michael Nielsen 著，地址：http://neuralnetworksanddeeplearning.com/。

Andrej Karpathy 用程序的方式介绍神经网络算法，其 JavaScript 例子称为"*A Hacker's Guide to Neural Networks*"，地址：http://karpathy.github.io/neuralnets/。

Ian Goodfellow、Yoshua Bengio 和 Aaron Courville 总结的深度学习笔记《Deep Learning for Beginners》，地址：http://randomekek.github.io/deep/deeplearning.html。

6.2 用 TensorFlow 实现门函数

神经网络算法的基本概念之一是门操作。本节以乘法操作作为门操作开始，接着介绍嵌套的门操作。

6.2.1 开始

第一个实现的操作门是 $f(x) = a \cdot x$。为了优化该门操作，我们声明 a 输入作为变量，x 输入作为占位符。这意味着 TensorFlow 将改变 a 的值，而不是 x 的值。我们将创建损失函数,度量输出结果和目标值之间的差值,这里的目标值是50。

第二个实现的嵌套操作门是 $f(x) = a \cdot x + b$。我们声明 a 和 b 为变量,x 为占位符。向目标值50优化输出结果。有趣的是第二个例子的解决方法不是唯一的。许多模型变量的组合使得输出结果为50。在神经网络算法中,我们不太关心模型变量的中间值,而把关注点放在预期的输出结果上。

想象一下,计算图中的操作门,图6-1是上述两个例子的门操作的描述。

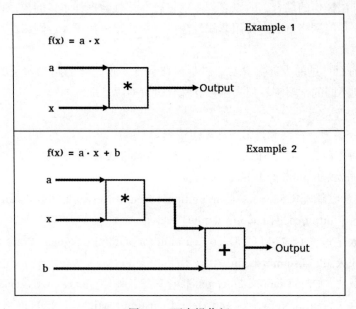

图6-1 两个操作门

6.2.2 动手做

为了在TensorFlow中实现第一个门操作 $f(x) = a \cdot x$,并训练输出结果50,具体的步骤如下。

1. 加载TensorFlow模块,创建一个计算图会话,代码如下:

```
import tensorflow as tf
sess = tf.Session()
```

2. 声明模型变量、输入数据集和占位符。本例输入数据为5,所以乘法因子为10,可以得到50的预期值($5 \times 10 = 50$),代码如下:

```
a = tf.Variable(tf.constant(4.))
x_val = 5.
x_data = tf.placeholder(dtype=tf.float32)
```

3. 增加操作到计算图中,代码如下:

```
multiplication = tf.multiply(a, x_data)
```

4. 声明损失函数：输出结果与预期目标值（50）之间的 L2 距离函数，代码如下：

```
loss = tf.square(tf.subtract(multiplication, 50.))
```

5. 初始化模型变量，声明标准梯度下降优化算法，代码如下：

```
init = tf.global_variables_initializer()
sess.run(init)
my_opt = tf.train.GradientDescentOptimizer(0.01)
train_step = my_opt.minimize(loss)
```

6. 优化模型输出结果，从而接近预期的输出值。连续输入值 5，反向传播损失函数来更新模型变量以达到值 10，代码如下：

```
print('Optimizing a Multiplication Gate Output to 50.')
for i in range(10):
    sess.run(train_step, feed_dict={x_data: x_val})
    a_val = sess.run(a)
    mult_output = sess.run(multiplication, feed_dict={x_data: x_val})
    print(str(a_val) + ' * ' + str(x_val) + ' = ' + str(mult_output))
```

7. 输出结果如下：

```
Optimizing a Multiplication Gate Output to 50.
7.0 * 5.0 = 35.0
8.5 * 5.0 = 42.5
9.25 * 5.0 = 46.25
9.625 * 5.0 = 48.125
9.8125 * 5.0 = 49.0625
9.90625 * 5.0 = 49.5312
9.95312 * 5.0 = 49.7656
9.97656 * 5.0 = 49.8828
9.98828 * 5.0 = 49.9414
9.99414 * 5.0 = 49.9707
```

对两个嵌套操作的例子 $f(x) = a \cdot x + b$，也执行上述相同的步骤。

8. 开始第二个例子，不同在于本例中包含两个模型变量：a 和 b，代码如下：

```
from tensorflow.python.framework import ops
ops.reset_default_graph()
sess = tf.Session()

a = tf.Variable(tf.constant(1.))
b = tf.Variable(tf.constant(1.))
x_val = 5.
x_data = tf.placeholder(dtype=tf.float32)

two_gate = tf.add(tf.multiply(a, x_data), b)

loss = tf.square(tf.subtract(two_gate, 50.))

my_opt = tf.train.GradientDescentOptimizer(0.01)
train_step = my_opt.minimize(loss)

init = tf.global_variables_initializer()
sess.run(init)
```

9. 优化模型变量，训练输出结果，以达到预期目标值50，代码如下：

```
print('Optimizing Two Gate Output to 50.')
for i in range(10):
    # Run the train step
    sess.run(train_step, feed_dict={x_data: x_val})
    # Get the a and b values
    a_val, b_val = (sess.run(a), sess.run(b))
    # Run the two-gate graph output
    two_gate_output = sess.run(two_gate, feed_dict={x_data: x_val})
    print(str(a_val) + ' * ' + str(x_val) + ' + ' + str(b_val) + ' = ' + str(two_gate_output))
```

10. 输出结果如下：

```
Optimizing Two Gate Output to 50.
5.4 * 5.0 + 1.88 = 28.88
7.512 * 5.0 + 2.3024 = 39.8524
8.52576 * 5.0 + 2.50515 = 45.134
9.01236 * 5.0 + 2.60247 = 47.6643
9.24593 * 5.0 + 2.64919 = 48.8789
9.35805 * 5.0 + 2.67161 = 49.4619
9.41186 * 5.0 + 2.68237 = 49.7417
9.43769 * 5.0 + 2.68754 = 49.876
9.45009 * 5.0 + 2.69002 = 49.9405
9.45605 * 5.0 + 2.69121 = 49.9714
```

> 这里需要注意的是，第二个例子的解决方法不是唯一的。这在神经网络算法中不太重要，因为所有的参数是根据减小损失函数来调整的。最终的解决方案依赖于a和b的初始值。如果它们是随机初始化的，而不是1，我们将会看到每次迭代的模型变量的输出结果并不相同。

6.2.3 工作原理

通过TensorFlow的隐式后向传播达到计算门操作的优化。TensorFlow维护模型操作和变量，调整优化算法和损失函数。

我们能扩展操作门，选定哪一个输入是变量，哪一个输入是数据。因为TensorFlow将调整所有的模型变量来最小化损失函数，而不是调整数据，数据输入声明为占位符。

维护计算图中的状态，以及每次训练迭代自动更新模型变量的隐式能力是TensorFlow具有的优势特征之一，该能力让TensorFlow威力无穷。

6.3 使用门函数和激励函数

现在我们已经学会了链接这些操作门函数，接下来将使用激励函数来运行计算图输出结果。本节会介绍常用的激励函数。

6.3.1 开始

本节将比较两种不同的激励函数：sigmoid 激励函数和 ReLU 激励函数。简单回忆一下这两个激励函数的表达式：

$$\text{sigmoid}(x) = \frac{1}{1+e^x}$$
$$\text{Re}LU(x) = \max(0, x)$$

在本例中，我们将创建两个相同结构的单层神经网络，有一点不同的是：一个通过 sigmoid 激励函数赋值；另外一个则通过 ReLU 激励函数赋值。损失函数使用 L2 范数距离函数（输出结果与 0.75 的差值）。我们将从正态分布数据集（Normal(mean = 2, sd = 0.1)）中随机抽取批量数据，优化输出结果达到预期值 0.75。

6.3.2 动手做

1. 导入必要的编程库，初始化一个计算图会话。对于学习在 TensorFlow 中如何设置随机种子而言，这也是一个很好的例子。这里将使用 TensorFlow 和 Numpy 模块的随机数生成器。对于相同的随机种子集，我们应该能够复现，代码如下：

```
import tensorflow as tf
import numpy as np
import matplotlib.pyplot as plt
sess = tf.Session()
tf.set_random_seed(5)
np.random.seed(42)
```

2. 声明批量大小、模型变量、数据集和占位符。在计算图中为两个相似的神经网络模型（仅激励函数不同）传入正态分布数据，代码如下：

```
batch_size = 50
a1 = tf.Variable(tf.random_normal(shape=[1,1]))
b1 = tf.Variable(tf.random_uniform(shape=[1,1]))
a2 = tf.Variable(tf.random_normal(shape=[1,1]))
b2 = tf.Variable(tf.random_uniform(shape=[1,1]))
x = np.random.normal(2, 0.1, 500)
x_data = tf.placeholder(shape=[None, 1], dtype=tf.float32)
```

3. 声明两个训练模型，即 sigmoid 激励模型和 ReLU 激励模型，代码如下：

```
sigmoid_activation = tf.sigmoid(tf.add(tf.matmul(x_data, a1), b1))
relu_activation = tf.nn.relu(tf.add(tf.matmul(x_data, a2), b2))
```

4. 损失函数都采用模型输出和预期值 0.75 之间的差值的 L2 范数平均，代码如下：

```
loss1 = tf.reduce_mean(tf.square(tf.subtract(sigmoid_activation, 0.75)))
loss2 = tf.reduce_mean(tf.square(tf.subtract(relu_activation, 0.75)))
```

5. 声明优化算法，初始化变量，代码如下：

```
my_opt = tf.train.GradientDescentOptimizer(0.01)
train_step_sigmoid = my_opt.minimize(loss1)
```

```
train_step_relu = my_opt.minimize(loss2)
init = tf.global_variable_initializer()
sess.run(init)
```

6. 遍历迭代训练模型，每个模型迭代 750 次。保存损失函数输出和激励函数的返回值，以便后续绘图，代码如下：

```
loss_vec_sigmoid = []
loss_vec_relu = []
activation_sigmoid = []
activation_relu = []
for i in range(750):
    rand_indices = np.random.choice(len(x), size=batch_size)
    x_vals = np.transpose([x[rand_indices]])
    sess.run(train_step_sigmoid, feed_dict={x_data: x_vals})
    sess.run(train_step_relu, feed_dict={x_data: x_vals})

    loss_vec_sigmoid.append(sess.run(loss1, feed_dict={x_data: x_vals}))
    loss_vec_relu.append(sess.run(loss2, feed_dict={x_data: x_vals}))
    activation_sigmoid.append(np.mean(sess.run(sigmoid_activation, feed_dict={x_data: x_vals})))
    activation_relu.append(np.mean(sess.run(relu_activation, feed_dict={x_data: x_vals})))
```

7. 下面是绘制损失函数和激励函数的代码，所绘图像见图 6-2：

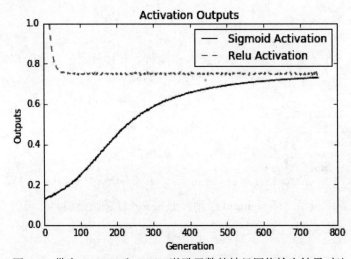

图 6-2　带有 sigmoid 和 ReLU 激励函数的神经网络输出结果对比

```
plt.plot(activation_sigmoid, 'k-', label='Sigmoid Activation')
plt.plot(activation_relu, 'r--', label='Relu Activation')
plt.ylim([0, 1.0])
plt.title('Activation Outputs')
plt.xlabel('Generation')
plt.ylabel('Outputs')
plt.legend(loc='upper right')
plt.show()
```

```
plt.plot(loss_vec_sigmoid, 'k-', label='Sigmoid Loss')
plt.plot(loss_vec_relu, 'r--', label='Relu Loss')
plt.ylim([0, 1.0])
plt.title('Loss per Generation')
plt.xlabel('Generation')
plt.ylabel('Loss')
plt.legend(loc='upper right')
plt.show()
```

两个神经网络模型具有相似的结构和预期值 0.75，只有激励函数不同（分别是 sigmoid 和 ReLU 激励函数）。从图 6-3 中可以看出，带有 ReLU 激励函数的神经网络比 sigmoid 激励函数的神经网络向 0.75 收敛得更快。

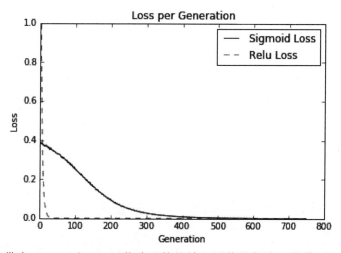

图 6-3　带有 sigmoid 和 ReLU 激励函数的神经网络的损失函数值对比。注意 ReLU 损失在迭代初期急剧下降

6.3.3　工作原理

基于 ReLU 激励函数的形式，它将比 sigmoid 激励函数返回更多的 0 值。我们认为该行为是一种稀疏性的，稀疏性导致收敛速度加快，但是损失了一部分梯度控制的能力。相反，sigmoid 激励函数具有良好的梯度控制，不会出现 ReLU 激励函数那样的极值。sigmoid 激励函数和 ReLU 激励函数的对比如下表所示。

激励函数	优点	缺点
Sigmoid 激励函数	输出的极值很少	收敛太慢
ReLU 激励函数	快速收敛	返回结果中容易出现极值

6.3.4　延伸学习

在本节中，我们比较了神经网络算法的 ReLU 激励函数和 sigmoid 激励函数。神经网络算法常用的其他激励函数还有很多,但是整体可以归纳成两类:第一类是形状类似 sigmoid 的激励函数,包括反正切激励函数、双曲正切函数、阶跃激励函数等;第二类是形状类似 ReLU 的激励函数,包括 softplus 激励函数、leak ReLU 激励函数等。本节讨论的两类激励函数的大部分比较结果也适用于所在的同类激励函数。然而,注意激励函数的选择对于神经网络算法的收敛是非常关键的。

6.4 用 TensorFlow 实现单层神经网络

我们已经实现了神经网络算法应用到真实数据集上的大部分操作,本节将实现一个单层神经网络(层即为神经网络中的神经元),并在 Iris 数据集上进行模型训练。

6.4.1 开始

在本节中,我们将实现一个单隐藏层的神经网络算法。理解全联接神经网络算法主要是基于矩阵乘法的,这一点是相当重要的。并且,数据集和矩阵的维度对于算法模型正确有序地运行是非常关键的。

由于本例是一个回归算法问题,所以将使用均方误差作为损失函数。

6.4.2 动手做

1. 创建计算图会话,导入必要的编程库,代码如下:

```
import matplotlib.pyplot as plt
import numpy as np
import tensorflow as tf
from sklearn import datasets
```

2. 加载 Iris 数据集,存储花萼长度作为目标值,然后开始一个计算图会话,代码如下:

```
iris = datasets.load_iris()
x_vals = np.array([x[0:3] for x in iris.data])
y_vals = np.array([x[3] for x in iris.data])
sess = tf.Session()
```

3. 因为数据集比较小,我们设置一个种子使得返回结果可复现,代码如下:

```
seed = 2
tf.set_random_seed(seed)
np.random.seed(seed)
```

4. 为了准备数据集,我们创建一个 80-20 分的训练集和测试集。通过 min-max 缩放法将 x 特征值归一化为 0～1 之间,代码如下:

```
train_indices = np.random.choice(len(x_vals),
round(len(x_vals)*0.8), replace=False)
test_indices = np.array(list(set(range(len(x_vals))) -
set(train_indices)))
x_vals_train = x_vals[train_indices]
x_vals_test = x_vals[test_indices]
```

```
y_vals_train = y_vals[train_indices]
y_vals_test = y_vals[test_indices]

def normalize_cols(m):
    col_max = m.max(axis=0)
    col_min = m.min(axis=0)
    return (m-col_min) / (col_max - col_min)

x_vals_train = np.nan_to_num(normalize_cols(x_vals_train))
x_vals_test = np.nan_to_num(normalize_cols(x_vals_test))
```

5. 现在为数据集和目标值声明批量大小和占位符，代码如下：

```
batch_size = 50
x_data = tf.placeholder(shape=[None, 3], dtype=tf.float32)
y_target = tf.placeholder(shape=[None, 1], dtype=tf.float32)
```

6. 这一步相当重要，声明有合适形状的模型变量。我们能声明隐藏层为任意大小，本例中设置为有五个隐藏节点，代码如下：

```
hidden_layer_nodes = 5
A1 = tf.Variable(tf.random_normal(shape=[3,hidden_layer_nodes]))
b1 = tf.Variable(tf.random_normal(shape=[hidden_layer_nodes]))
A2 = tf.Variable(tf.random_normal(shape=[hidden_layer_nodes,1]))
b2 = tf.Variable(tf.random_normal(shape=[1]))
```

7. 分两步声明训练模型：第一步，创建一个隐藏层输出；第二步，创建训练模型的最后输出，代码如下：

 注意，本例中的模型有三个特征、五个隐藏节点和一个输出结果值。

```
hidden_output = tf.nn.relu(tf.add(tf.matmul(x_data, A1), b1))
final_output = tf.nn.relu(tf.add(tf.matmul(hidden_output, A2), b2))
```

8. 这里定义均方误差作为损失函数，代码如下：

```
loss = tf.reduce_mean(tf.square(y_target - final_output))
```

9. 声明优化算法，初始化模型变量，代码如下：

```
my_opt = tf.train.GradientDescentOptimizer(0.005)
train_step = my_opt.minimize(loss)
init = tf.global_variables_initializer()
sess.run(init)
```

10. 遍历迭代训练模型。我们也初始化两个列表（list）存储训练损失和测试损失。在每次迭代训练时，随机选择批量训练数据来拟合模型，代码如下：

```
# First we initialize the loss vectors for storage.
loss_vec = []
test_loss = []
for i in range(500):
    # We select a random set of indices for the batch.
    rand_index = np.random.choice(len(x_vals_train),
```

```
size=batch_size)
    # We then select the training values
    rand_x = x_vals_train[rand_index]
    rand_y = np.transpose([y_vals_train[rand_index]])
    # Now we run the training step
    sess.run(train_step, feed_dict={x_data: rand_x, y_target:
rand_y})
    # We save the training loss
    temp_loss = sess.run(loss, feed_dict={x_data: rand_x, y_target:
rand_y})
    loss_vec.append(np.sqrt(temp_loss))

    # Finally, we run the test-set loss and save it.
    test_temp_loss = sess.run(loss, feed_dict={x_data: x_vals_test,
y_target: np.transpose([y_vals_test])})
    test_loss.append(np.sqrt(test_temp_loss))
    if (i+1)%50==0:
        print('Generation: ' + str(i+1) + '. Loss = ' +
str(temp_loss))
```

11. 使用 matplotlib 绘制损失函数的代码如下，所绘图像如图 6-4 所示：

```
plt.plot(loss_vec, 'k-', label='Train Loss')
plt.plot(test_loss, 'r--', label='Test Loss')
plt.title('Loss (MSE) per Generation')
plt.xlabel('Generation')
plt.ylabel('Loss')
plt.legend(loc='upper right')
plt.show()
```

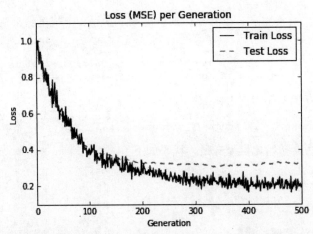

图 6-4　训练集和测试集的损失函数（MSE）绘图。注意，在 200 次迭代训练后会出现轻微的过拟合，因为测试集 MSE 没有丢失特征，但是训练集 MSE 会剔除特征

6.4.3　工作原理

可视化神经网络算法模型，如图 6-5 所示。

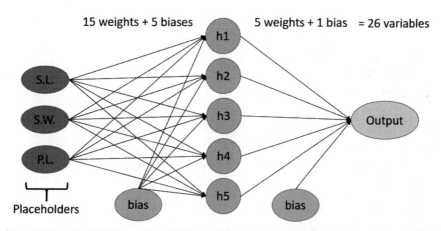

图 6-5 神经网络算法模型可视化，其中隐藏层包含五个隐藏节点。我们传入三个值：花萼长度（S.L）、花萼宽度（S.W）和花瓣长度（P.L）。目标值是花瓣宽度。总共将有 26 个模型变量

6.4.4 延伸学习

注意，我们通过可视化测试集和训练集的损失函数可以判定训练数据集上的模型训练是否过拟合，也可以发现测试集的损失函数比训练集更平滑，主要有两个原因：第一个是，训练集的数据批量大小比测试集小（虽然小得不太多）；第二个是，模型训练是在训练集上进行的，所以测试集对模型变量没有影响。

6.5 用 TensorFlow 实现神经网络常见层

本节将介绍如何实现常见层，包括卷积层（convolutional layer）和池化层（maxpool layer）。在上一节，我们实现了全联接层，本小节将扩展到其他层。

6.5.1 开始

我们已经研究了如何连接数据输入和全联接的隐藏层。TensorFlow 中有许多内建函数的多种类型的层，其中最流行的层是卷积层和池化层。我们将展示如何在输入数据和全联接的数据上创建和使用这些层。首先，我们来介绍如何在一维数据上使用这些层，然后是在二维数据上使用它们。

神经网络算法的层能以任意形式组合，最常用的使用方法是用卷积层和全联接层来创建特征。如果我们有许多特征，常用的处理方法是采用池化层。在这些层之后常常引入激励函数。卷积神经网络（CNN）算法（见第 8 章），包括卷积层、池化层、激励函数等。

6.5.2 动手做

刚开始，先以一维数据为例。我们将生成一维随机数据。

1. 导入需要的编程库，创建计算图会话，代码如下：

```
import tensorflow as tf
import numpy as np
sess = tf.Session()
```

2. 初始化数据，该数据为 NumPy 数组，长度为 25。创建传入数据的占位符，代码如下：

```
data_size = 25
data_1d = np.random.normal(size=data_size)
x_input_1d = tf.placeholder(dtype=tf.float32, shape=[data_size])
```

3. 定义一个卷积层的函数。接着声明一个随机过滤层，创建一个卷积层，代码如下：

> 注意，许多 TensorFlow 的层函数是为四维数据设计的（4D = [batch size, width, height, channels]）。我们需要调整输入数据和输出数据，包括扩展维度和降维。在本例中，批量大小为 1，宽度为 1，高度为 25，颜色通道为 1。为了扩展维度，使用 expand_dims() 函数；降维使用 squeeze() 函数。卷积层的输出结果的维度公式为 output_size = (W - F + 2P)/S + 1，其中 W 为输入数据维度，F 为过滤层大小，P 是 padding 大小，S 是步长大小。

```
def conv_layer_1d(input_1d, my_filter):
    # Make 1d input into 4d
    input_2d = tf.expand_dims(input_1d, 0)
    input_3d = tf.expand_dims(input_2d, 0)
    input_4d = tf.expand_dims(input_3d, 3)
    # Perform convolution
    convolution_output = tf.nn.conv2d(input_4d, filter=my_filter, strides=[1,1,1,1], padding="VALID")
    # Now drop extra dimensions
    conv_output_1d = tf.squeeze(convolution_output)
    return(conv_output_1d)

my_filter = tf.Variable(tf.random_normal(shape=[1,5,1,1]))
my_convolution_output = conv_layer_1d(x_input_1d, my_filter)
```

4. TensorFlow 的激励函数默认是逐个元素进行操作。这意味着，在部分层中需要使用激励函数。下面创建一个激励函数并初始化：

```
def activation(input_1d):
    return tf.nn.relu(input_1d)
my_activation_output = activation(my_convolution_output)
```

5. 声明一个池化层函数，该函数在一维向量的移动窗口上创建池化层函数。对于本例，其宽度为 5，代码如下：

> TensorFlow 的池化层函数的参数与卷积层函数参数非常相似。但是它没有过滤层，只有形状、步长和 padding 选项。因为我们的窗口宽度为 5，并且具有 valid padding（即非零 padding），所以输出数组将有 4 或者 2·floor(5/2) 项。

```
def max_pool(input_1d, width):
    # First we make the 1d input into 4d.
    input_2d = tf.expand_dims(input_1d, 0)
    input_3d = tf.expand_dims(input_2d, 0)
    input_4d = tf.expand_dims(input_3d, 3)
    # Perform the max pool operation
    pool_output = tf.nn.max_pool(input_4d, ksize=[1, 1, width, 1],
strides=[1, 1, 1, 1], padding='VALID')
    pool_output_1d = tf.squeeze(pool_output)
    return pool_output_1d

my_maxpool_output = max_pool(my_activation_output, width=5)
```

6. 最后一层连接的是全联接层。创建一个函数，该函数输入一维数据，输出值的索引。记住一维数组做矩阵乘法需要提前扩展为二维，代码如下：

```
def fully_connected(input_layer, num_outputs):
    # Create weights
    weight_shape = tf.squeeze(tf.stack([tf.shape(input_layer),
[num_outputs]]))
    weight = tf.random_normal(weight_shape, stddev=0.1)
    bias = tf.random_normal(shape=[num_outputs])
    # Make input into 2d
    input_layer_2d = tf.expand_dims(input_layer, 0)
    # Perform fully connected operations
    full_output = tf.add(tf.matmul(input_layer_2d, weight), bias)
    # Drop extra dimensions
    full_output_1d = tf.squeeze(full_output)
    return full_output_1d

my_full_output = fully_connected(my_maxpool_output, 5)
```

7. 初始化所有的变量，运行计算图打印出每层的输出结果，代码如下：

```
init = tf.global_variable_initializer()
sess.run(init)
feed_dict = {x_input_1d: data_1d}
# Convolution Output
print('Input = array of length 25')
print('Convolution w/filter, length = 5, stride size = 1, results
in an array of length 21:')
print(sess.run(my_convolution_output, feed_dict=feed_dict))
# Activation Output
print('Input = the above array of length 21')
print('ReLU element wise returns the array of length 21:')
print(sess.run(my_activation_output, feed_dict=feed_dict))
# Maxpool Output
print('Input = the above array of length 21')
print('MaxPool, window length = 5, stride size = 1, results in the
array of length 17:')
print(sess.run(my_maxpool_output, feed_dict=feed_dict))
# Fully Connected Output
print('Input = the above array of length 17')
print('Fully connected layer on all four rows with five outputs:')
print(sess.run(my_full_output, feed_dict=feed_dict))
```

8. 输出结果如下:

```
Input = array of length 25
Convolution w/filter, length = 5, stride size = 1, results in an
array of length 21:
[-0.91608119  1.53731811 -0.7954089   0.5041104   1.88933098
 -1.81099761  0.56695032  1.17945457 -0.66252393 -1.90287709
  0.87184119  0.84611893 -5.25024986 -0.05473572  2.19293165
 -4.47577858 -1.71364677  3.96857905 -2.0452652  -1.86647367
 -0.12697852]
Input = the above array of length 21
ReLU element wise returns the array of length 21:
[ 0.          1.53731811  0.          0.5041104   1.88933098
  0.          0.          1.17945457  0.          0.
  0.87184119  0.84611893  0.          0.          2.19293165
  0.          0.          3.96857905  0.          0.
  0.         ]
Input = the above array of length 21
MaxPool, window length = 5, stride size = 1, results in the array
of length 17:
[ 1.88933098  1.88933098  1.88933098  1.88933098  1.88933098
  1.17945457  1.17945457  1.17945457  0.87184119  0.87184119
  2.19293165  2.19293165  2.19293165  3.96857905  3.96857905
  3.96857905  3.96857905]
Input = the above array of length 17
Fully connected layer on all four rows with five outputs:
[ 1.23588216 -0.42116445  1.44521213  1.40348077 -0.79607368]
```

> 神经网络对于一维数据非常重要。时序数据集、信号处理数据集和一些文本嵌入数据集都是一维数据,会频繁使用到神经网络算法。

下面开始在二维数据集上进行层函数操作:

1. 重置计算图会话,代码如下:

```
ops.reset_default_graph()
sess = tf.Session()
```

2. 初始化输入数组为 10×10 的矩阵,然后初始化计算图的占位符,代码如下:

```
data_size = [10,10]
data_2d = np.random.normal(size=data_size)
x_input_2d = tf.placeholder(dtype=tf.float32, shape=data_size)
```

3. 声明一个卷积层函数。因为数据集已经具有高度和宽度了,这里仅需再扩展两维(批量大小为 1,颜色通道为 1)即可使用卷积 conv2d() 函数。本例将使用一个随机的 2×2 过滤层,两个方向上的步长和 valid padding (非零 padding)。由于输入数据是 10×10,因此卷积输出为 5×5。具体代码如下:

```
def conv_layer_2d(input_2d, my_filter):
    # First, change 2d input to 4d
    input_3d = tf.expand_dims(input_2d, 0)
    input_4d = tf.expand_dims(input_3d, 3)
    # Perform convolution
```

```
        convolution_output = tf.nn.conv2d(input_4d, filter=my_filter,
        strides=[1,2,2,1], padding="VALID")
        # Drop extra dimensions
        conv_output_2d = tf.squeeze(convolution_output)
        return(conv_output_2d)

my_filter = tf.Variable(tf.random_normal(shape=[2,2,1,1]))
my_convolution_output = conv_layer_2d(x_input_2d, my_filter)
```

4. 激励函数是针对逐个元素的，现创建激励函数并初始化，代码如下：

```
def activation(input_2d):
    return tf.nn.relu(input_2d)
my_activation_output = activation(my_convolution_output)
```

5. 本例的池化层与一维数据例子中的相似，有一点不同的是，我们需要声明池化层移动窗口的宽度和高度。这里将与二维卷积层一样，将扩展池化层为二维，代码如下：

```
def max_pool(input_2d, width, height):
    # Make 2d input into 4d
    input_3d = tf.expand_dims(input_2d, 0)
    input_4d = tf.expand_dims(input_3d, 3)
    # Perform max pool
    pool_output = tf.nn.max_pool(input_4d, ksize=[1, height, width,
1], strides=[1, 1, 1, 1], padding='VALID')
    # Drop extra dimensions
    pool_output_2d = tf.squeeze(pool_output)
    return pool_output_2d

my_maxpool_output = max_pool(my_activation_output, width=2,
height=2)
```

6. 本例中的全联接层也与一维数据的输出相似。注意，全联接层的二维输入看作一个对象，为了实现每项连接到每个输出，我们展开二维矩阵，然后在做矩阵乘法时再扩展维度，代码如下：

```
def fully_connected(input_layer, num_outputs):
    # Flatten into 1d
    flat_input = tf.reshape(input_layer, [-1])
    # Create weights
    weight_shape = tf.squeeze(tf.stack([tf.shape(flat_input),
[num_outputs]]))
    weight = tf.random_normal(weight_shape, stddev=0.1)
    bias = tf.random_normal(shape=[num_outputs])
    # Change into 2d
    input_2d = tf.expand_dims(flat_input, 0)
    # Perform fully connected operations
    full_output = tf.add(tf.matmul(input_2d, weight), bias)
    # Drop extra dimensions
    full_output_2d = tf.squeeze(full_output)
    return full_output_2d
my_full_output = fully_connected(my_maxpool_output, 5)
```

7. 初始化变量，创建一个赋值字典，代码如下：

```
init = tf.global_variables_initializer()
```

```
sess.run(init)

feed_dict = {x_input_2d: data_2d}
```

8. 打印每层的输出结果，代码如下：

```
# Convolution Output
print('Input = [10 X 10] array')
print('2x2 Convolution, stride size = [2x2], results in the [5x5] array:')
print(sess.run(my_convolution_output, feed_dict=feed_dict))
# Activation Output
print('Input = the above [5x5] array')
print('ReLU element wise returns the [5x5] array:')
print(sess.run(my_activation_output, feed_dict=feed_dict))
# Max Pool Output
print('Input = the above [5x5] array')
print('MaxPool, stride size = [1x1], results in the [4x4] array:')
print(sess.run(my_maxpool_output, feed_dict=feed_dict))
# Fully Connected Output
print('Input = the above [4x4] array')
print('Fully connected layer on all four rows with five outputs:')
print(sess.run(my_full_output, feed_dict=feed_dict))
```

9. 输出结果如下：

```
Input = [10 X 10] array
2x2 Convolution, stride size = [2x2], results in the [5x5] array:
[[ 0.37630892 -1.41018617 -2.58821273 -0.32302785  1.18970704]
 [-4.33685207  1.97415686  1.0844903  -1.18965471  0.84643292]
 [ 5.23706436  2.46556497 -0.95119286  1.17715418  4.1117816 ]
 [ 5.86972761  1.2213701   1.59536231  2.66231227  2.28650784]
 [-0.88964868 -2.75502229  4.3449688   2.67776585 -2.23714781]]
Input = the above [5x5] array
ReLU element wise returns the [5x5] array:
[[ 0.37630892  0.          0.          0.          1.18970704]
 [ 0.          1.97415686  1.0844903   0.          0.84643292]
 [ 5.23706436  2.46556497  0.          1.17715418  4.1117816 ]
 [ 5.86972761  1.2213701   1.59536231  2.66231227  2.28650784]
 [ 0.          0.          4.3449688   2.67776585  0.        ]]
Input = the above [5x5] array
MaxPool, stride size = [1x1], results in the [4x4] array:
[[ 1.97415686  1.97415686  1.0844903   1.18970704]
 [ 5.23706436  2.46556497  1.17715418  4.1117816 ]
 [ 5.86972761  2.46556497  2.66231227  4.1117816 ]
 [ 5.86972761  4.3449688   4.3449688   2.67776585]]
Input = the above [4x4] array
Fully connected layer on all four rows with five outputs:
[-0.6154139  -1.96987963 -1.88811922  0.20010889  0.32519674]
```

6.5.3 工作原理

我们学习了如何在一维数据集和二维数据集上使用 TensorFlow 的卷积层和池化层。不管输入数据集的形状，最后的输出结果都是相同维度的。这现实了神经网络算法层的灵活性。本节也让我们理解了，神经网络操作中形状和大小的重要性。

6.6 用 TensorFlow 实现多层神经网络

本节将多层神经网络应用到实际场景中，预测低出生体重数据集。

6.6.1 开始

截至目前，我们学习了如何创建神经网络和层，我们将运用该方法在低出生体重数据集上预测婴儿出生体重。我们将创建一个包含三个隐藏层的神经网络。低出生体重数据集包括实际的出生体重和是否超过 2500 克的标记。在本例中，我们将预测出生体重（回归预测），然后看最后分类结果的准确度（模型能够鉴别出生体重是否超过 2500 克）。

6.6.2 动手做

1. 导入必要的编程库，初始化计算图会话，代码如下：

```
import tensorflow as tf
import matplotlib.pyplot as plt
import os
import csv
import requests
import numpy as np
sess = tf.Session()
```

2. 使用 requests 模块从网站加载数据集，然后分离出需要的特征数据和目标值，代码如下：

```
# Name of data file
birth_weight_file = 'birth_weight.csv'
birthdata_url = 'https://github.com/nfmcclure/tensorflow_cookbook/raw/master' \
'/01_Introduction/07_Working_with_Data_Sources/birthweight_data/birthweight.dat'

# Download data and create data file if file does not exist in current directory
if not os.path.exists(birth_weight_file):
    birth_file = requests.get(birthdata_url)
    birth_data = birth_file.text.split('\r\n')
    birth_header = birth_data[0].split('\t')
    birth_data = [[float(x) for x in y.split('\t') if len(x) >= 1]
                  for y in birth_data[1:] if len(y) >= 1]
    with open(birth_weight_file, "w") as f:
        writer = csv.writer(f)
        writer.writerows([birth_header])
        writer.writerows(birth_data)

# Read birth weight data into memory
birth_data = []
with open(birth_weight_file, newline='') as csvfile:
    csv_reader = csv.reader(csvfile)
    birth_header = next(csv_reader)
    for row in csv_reader:
        birth_data.append(row)
birth_data = [[float(x) for x in row] for row in birth_data]
```

```
# Pull out target variable
y_vals = np.array([x[0] for x in birth_data])
# Pull out predictor variables (not id, not target, and not
birthweight)
x_vals = np.array([x[1:8] for x in birth_data])
```

3. 为了后面可以复现，为 NumPy 和 TensorFlow 设置随机种子，然后声明批量大小，代码如下：

```
seed = 4
tf.set_random_seed(seed)
np.random.seed(seed)
batch_size = 100
```

4. 分割数据集为 80-20 的训练集和测试集，然后使用 min-max 方法归一化输入特征数据为 0 到 1 之间，代码如下：

```
train_indices = np.random.choice(len(x_vals),
round(len(x_vals)*0.8), replace=False)
test_indices = np.array(list(set(range(len(x_vals))) -
set(train_indices)))
x_vals_train = x_vals[train_indices]
x_vals_test = x_vals[test_indices]
y_vals_train = y_vals[train_indices]
y_vals_test = y_vals[test_indices]

# Normalize by column (min-max norm)
def normalize_cols(m, col_min=np.array([None]),
col_max=np.array([None])):
    if not col_min[0]:
        col_min = m.min(axis=0)
    if not col_max[0]:
        col_max = m.max(axis=0)
    return (m-col_min) / (col_max - col_min), col_min, col_max

x_vals_train, train_min, train_max =
np.nan_to_num(normalize_cols(x_vals_train))
x_vals_test, _, _ = np.nan_to_num(normalize_cols(x_vals_test),
train_min, train_max)
```

> 归一化输入特征数据是常用的特征转化方法，对神经网络算法特别有帮助。如果样本数据集是以 0 到 1 为中心的，它将有利于激励函数操作的收敛。

5. 因为有多个层含有相似的变量初始化，因此我们将创建一个初始化函数，该函数可以初始化加权权重和偏置，代码如下：

```
def init_weight(shape, st_dev):
    weight = tf.Variable(tf.random_normal(shape, stddev=st_dev))
    return weight


def init_bias(shape, st_dev):
    bias = tf.Variable(tf.random_normal(shape, stddev=st_dev))
    return bias
```

6. 初始化占位符。本例中将有八个输入特征数据和一个输出结果（出生体重，单位：克），代码如下：

```
x_data = tf.placeholder(shape=[None, 8], dtype=tf.float32)
y_target = tf.placeholder(shape=[None, 1], dtype=tf.float32)
```

7. 全联接层将在三个隐藏层中使用三次，为了避免代码上的重复，我们将创建一个层函数来初始化算法模型，代码如下：

```
def fully_connected(input_layer, weights, biases):
    layer = tf.add(tf.matmul(input_layer, weights), biases)
    return tf.nn.relu(layer)
```

8. 现在创建算法模型。对于每一层（包括输出层），我们将初始化一个权重矩阵、偏置矩阵和全联接层。在本例中，三个隐藏层的大小分别为25、10和3，代码如下：

> 本例中使用的算法模型需要拟合522个变量。下面来看下这个数值是如何计算的？输入数据集和第一隐藏层之间有225（8×25 + 25）个变量，继续用这种方式计算隐藏层并加在一起有522（225 + 260 + 33 + 4）个变量。很明显，这比之前在逻辑回归算法中的9个变量要多得多。

```
# Create second layer (25 hidden nodes)
weight_1 = init_weight(shape=[8, 25], st_dev=10.0)
bias_1 = init_bias(shape=[25], st_dev=10.0)
layer_1 = fully_connected(x_data, weight_1, bias_1)
# Create second layer (10 hidden nodes)
weight_2 = init_weight(shape=[25, 10], st_dev=10.0)
bias_2 = init_bias(shape=[10], st_dev=10.0)
layer_2 = fully_connected(layer_1, weight_2, bias_2)

# Create third layer (3 hidden nodes)
weight_3 = init_weight(shape=[10, 3], st_dev=10.0)
bias_3 = init_bias(shape=[3], st_dev=10.0)
layer_3 = fully_connected(layer_2, weight_3, bias_3)
# Create output layer (1 output value)
weight_4 = init_weight(shape=[3, 1], st_dev=10.0)
bias_4 = init_bias(shape=[1], st_dev=10.0)
final_output = fully_connected(layer_3, weight_4, bias_4)
```

9. 使用L1范数损失函数（绝对值），声明优化器（Adam优化器）和初始化变量，代码如下：

```
loss = tf.reduce_mean(tf.abs(y_target - final_output))
my_opt = tf.train.AdamOptimizer(0.05)
train_step = my_opt.minimize(loss)
init = tf.global_variables_initializer()
sess.run(init)
```

> 本例中为Adam优化器选择的学习率为0.05，有研究建议设置更低的学习率可以产生更好的结果。在本节中，我们使用比较大的学习率是为了数据集的一致性和快速收敛。

10. 迭代训练模型 200 次。下面的代码也包括存储训练损失和测试损失，选择随机批量大小和每 25 次迭代就打印状态，代码如下：

```
# Initialize the loss vectors
loss_vec = []
test_loss = []
for i in range(200):
    # Choose random indices for batch selection
    rand_index = np.random.choice(len(x_vals_train), size=batch_size)
    # Get random batch
    rand_x = x_vals_train[rand_index]
    rand_y = np.transpose([y_vals_train[rand_index]])
    # Run the training step
    sess.run(train_step, feed_dict={x_data: rand_x, y_target: rand_y})
    # Get and store the train loss
    temp_loss = sess.run(loss, feed_dict={x_data: rand_x, y_target: rand_y})
    loss_vec.append(temp_loss)
    # Get and store the test loss
    test_temp_loss = sess.run(loss, feed_dict={x_data: x_vals_test, y_target: np.transpose([y_vals_test])})
    test_loss.append(test_temp_loss)
    if (i+1)%25==0:
        print('Generation: ' + str(i+1) + '. Loss = ' + str(temp_loss))
```

11. 输出结果如下：

```
Generation: 25. Loss = 5922.52
Generation: 50. Loss = 2861.66
Generation: 75. Loss = 2342.01
Generation: 100. Loss = 1880.59
Generation: 125. Loss = 1394.39
Generation: 150. Loss = 1062.43
Generation: 175. Loss = 834.641
Generation: 200. Loss = 848.54
```

12. 使用 matplotlib 模块绘制训练损失和测试损失的代码，所绘图像见图 6-6：

```
plt.plot(loss_vec, 'k-', label='Train Loss')
plt.plot(test_loss, 'r--', label='Test Loss')
plt.title('Loss per Generation')
plt.xlabel('Generation')
plt.ylabel('Loss')
plt.legend(loc='upper right')
plt.show()
```

13. 现在我们想比较预测出生体重结果和前面章节的逻辑结果。在第 3 章的逻辑回归算法中，我们在迭代上千次后得到了大约 60% 的精确度。为了在这里做比较，我们将输出训练集 / 测试集的回归结果，然后传入一个指示函数（判断是否大于 2500 克），将回归结果转换成分类结果。下面的代码将证明本例模型的准确度：

```
actuals = np.array([x[1] for x in birth_data])
test_actuals = actuals[test_indices]
```

```
train_actuals = actuals[train_indices]
test_preds = [x[0] for x in sess.run(final_output,
    feed_dict={x_data: x_vals_test})]
train_preds = [x[0] for x in sess.run(final_output,
    feed_dict={x_data: x_vals_train})]
test_preds = np.array([1.0 if x<2500.0 else 0.0 for x in
test_preds])
train_preds = np.array([1.0 if x<2500.0 else 0.0 for x in
train_preds])
# Print out accuracies
test_acc = np.mean([x==y for x,y in zip(test_preds, test_actuals)])
train_acc = np.mean([x==y for x,y in zip(train_preds,
train_actuals)])
print('On predicting the category of low birthweight from
regression output (<2500g):')
print('Test Accuracy: {}'.format(test_acc))
print('Train Accuracy: {}'.format(train_acc))
```

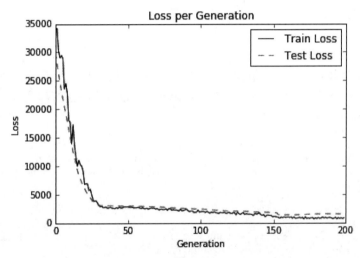

图 6-6 神经网络算法模型训练预测出生体重的训练集损失和测试集损失图（单位：克）。注意，在 30 次迭代训练后会获得较好的模型

14. 准确度的结果如下：

Test Accuracy: 0.631578947368421
Train Accuracy: 0.7019867549668874

6.6.3 工作原理

在本节中，我们创建一个回归神经网络模型，该模型有三个全联接隐藏层，并用该模型预测低出生体重数据集。与逻辑结果预测出生体重是否大于 2500 克相比，我们得到了相似的结果，并且迭代的次数更少。在下一节中，我们将通过多层逻辑神经网络算法来优化逻辑回归算法模型。

6.7 线性预测模型的优化

在上一节中,我们注意到需拟合的参数数量远超线性模型。在本节中,我们试图用神经网络算法模型来优化低出生体重的逻辑模型。

6.7.1 开始

加载低出生体重样本数据集,使用一个带两个隐藏层的全联接层的神经网络,并采用 sigmoid 激励函数来拟合低出生体重的概率。

6.7.2 动手做

1. 导入必要的编程库,初始化计算图会话,代码如下:

```
import matplotlib.pyplot as plt
import numpy as np
import tensorflow as tf
import requests
sess = tf.Session()
```

2. 加载低出生体重数据集,并对其进行抽取和归一化。有一点不同的是,本例中将使用低出生体重指示变量作为目标值,而不是实际出生体重,代码如下:

```
# Name of data file
birth_weight_file = 'birth_weight.csv'
birthdata_url = \
'https://github.com/nfmcclure/tensorflow_cookbook/raw/master' \
'/01_Introduction/07_Working_with_Data_Sources/birthweight_data/bir
thweight.dat'

# Download data and create data file if file does not exist in
current directory
if not os.path.exists(birth_weight_file):
    birth_file = requests.get(birthdata_url)
    birth_data = birth_file.text.split('\r\n')
    birth_header = birth_data[0].split('\t')
    birth_data = [[float(x) for x in y.split('\t') if len(x) >= 1]
                  for y in birth_data[1:] if len(y) >= 1]
    with open(birth_weight_file, "w") as f:
        writer = csv.writer(f)
        writer.writerows([birth_header])
        writer.writerows(birth_data)

# read birth weight data into memory
birth_data = []
with open(birth_weight_file, newline='') as csvfile:
    csv_reader = csv.reader(csvfile)
    birth_header = next(csv_reader)
    for row in csv_reader:
        birth_data.append(row)

birth_data = [[float(x) for x in row] for row in birth_data]
# Pull out target variable
```

```
y_vals = np.array([x[0] for x in birth_data])
# Pull out predictor variables (not id, not target, and not
birthweight)
x_vals = np.array([x[1:8] for x in birth_data])

train_indices = np.random.choice(len(x_vals),
round(len(x_vals)*0.8), replace=False)
test_indices = np.array(list(set(range(len(x_vals))) -
set(train_indices)))
x_vals_train = x_vals[train_indices]
x_vals_test = x_vals[test_indices]
y_vals_train = y_vals[train_indices]
y_vals_test = y_vals[test_indices]

def normalize_cols(m, col_min=np.array([None]),
col_max=np.array([None])):
    if not col_min[0]:
        col_min = m.min(axis=0)
    if not col_max[0]:
        col_max = m.max(axis=0)
    return (m - col_min) / (col_max - col_min), col_min, col_max


x_vals_train, train_min, train_max =
np.nan_to_num(normalize_cols(x_vals_train))
x_vals_test, _, _ = np.nan_to_num(normalize_cols(x_vals_test,
train_min, train_max))
```

3. 声明批量大小和占位符，代码如下：

```
batch_size = 90
x_data = tf.placeholder(shape=[None, 7], dtype=tf.float32)
y_target = tf.placeholder(shape=[None, 1], dtype=tf.float32)
```

4. 我们声明函数来初始化算法模型中的变量和层。为了创建一个更好的逻辑层，我们需要创建一个返回输入层的逻辑层的函数。换句话说，我们需要使用全联接层，返回每层的 sigmoid 值。注意，损失函数包括最终的 sigmoid 函数，所以我们指定最后一层不必返回输出的 sigmoid 值，代码如下：

```
def init_variable(shape):
    return tf.Variable(tf.random_normal(shape=shape))
# Create a logistic layer definition
def logistic(input_layer, multiplication_weight, bias_weight,
activation = True):
    linear_layer = tf.add(tf.matmul(input_layer,
multiplication_weight), bias_weight)

    if activation:
        return tf.nn.sigmoid(linear_layer)
    else:
        return linear_layer
```

5. 声明神经网络的三层（两个隐藏层和一个输出层）。我们为每层初始化一个权重矩阵和偏置矩阵，并定义每层的操作，代码如下：

```
# First logistic layer (7 inputs to 14 hidden nodes)
A1 = init_variable(shape=[7,14])
```

```
b1 = init_variable(shape=[14])
logistic_layer1 = logistic(x_data, A1, b1)

# Second logistic layer (14 hidden inputs to 5 hidden nodes)
A2 = init_variable(shape=[14,5])
b2 = init_variable(shape=[5])
logistic_layer2 = logistic(logistic_layer1, A2, b2)
# Final output layer (5 hidden nodes to 1 output)
A3 = init_variable(shape=[5,1])
b3 = init_variable(shape=[1])
final_output = logistic(logistic_layer2, A3, b3, activation=False)
```

6. 声明损失函数（本例使用的是交叉熵损失函数）和优化算法，并初始化变量，代码如下：

```
# Create loss function
loss = tf.reduce_mean(tf.nn.sigmoid_cross_entropy_with_logits(logits=final
_output, labels=y_target))
# Declare optimizer
my_opt = tf.train.AdamOptimizer(learning_rate = 0.002)
train_step = my_opt.minimize(loss)
# Initialize variables
init = tf.global_variables_initializer()
sess.run(init)
```

> 交叉熵是度量概率之间的距离。这里度量确定值（0或者1）和模型概率值（0< x <1）之间的差值。在 TensorFlow 中实现的交叉熵是用函数 sigmoid() 内建的。采用超参数调优对于寻找最好的损失函数、学习率和优化算法是相当重要的，但是为了本节示例的简洁性，这里不介绍超参数调优。

7. 为了评估和比较算法模型，创建计算图预测操作和准确度操作。这使得我们可以传入测试集并计算准确度，代码如下：

```
prediction = tf.round(tf.nn.sigmoid(final_output))
predictions_correct = tf.cast(tf.equal(prediction, y_target),
tf.float32)
accuracy = tf.reduce_mean(predictions_correct)
```

8. 准备开始遍历迭代训练模型。本例将训练1500次，并为后续绘图保存模型的损失函数和训练集/测试集准确度，代码如下：

```
# Initialize loss and accuracy vectors loss_vec = [] train_acc = []
test_acc = []
for i in range(1500):
    # Select random indicies for batch selection
    rand_index = np.random.choice(len(x_vals_train),
size=batch_size)
    # Select batch
    rand_x = x_vals_train[rand_index]
    rand_y = np.transpose([y_vals_train[rand_index]])
    # Run training step
    sess.run(train_step, feed_dict={x_data: rand_x, y_target:
```

```
rand_y})
    # Get training loss
    temp_loss = sess.run(loss, feed_dict={x_data: rand_x, y_target:
rand_y})
    loss_vec.append(temp_loss)
    # Get training accuracy
    temp_acc_train = sess.run(accuracy, feed_dict={x_data:
x_vals_train, y_target: np.transpose([y_vals_train])})
    train_acc.append(temp_acc_train)
    # Get test accuracy
    temp_acc_test = sess.run(accuracy, feed_dict={x_data:
x_vals_test, y_target: np.transpose([y_vals_test])})
    test_acc.append(temp_acc_test)
    if (i+1)%150==0:
        print('Loss = ' + str(temp_loss))
```

9. 输出结果如下：

Loss = 0.696393
Loss = 0.591708
Loss = 0.59214
Loss = 0.505553
Loss = 0.541974
Loss = 0.512707
Loss = 0.590149
Loss = 0.502641
Loss = 0.518047
Loss = 0.502616

10. 下面的代码块展示如何用matplotlib模块绘制交叉熵损失函数和测试集/训练集准确度：

```
# Plot loss over time
plt.plot(loss_vec, 'k-')
plt.title('Cross Entropy Loss per Generation')
plt.xlabel('Generation')
plt.ylabel('Cross Entropy Loss')
plt.show()
# Plot train and test accuracy
plt.plot(train_acc, 'k-', label='Train Set Accuracy')
plt.plot(test_acc, 'r--', label='Test Set Accuracy')
plt.title('Train and Test Accuracy')
plt.xlabel('Generation')
plt.ylabel('Accuracy')
plt.legend(loc='lower right')
plt.show()
```

从图6-7可以看出，大约迭代50次，我们得到了较好的训练模型。随着继续迭代训练，我们发现后续的迭代并没有获得较大的效果提升。

从图6-8可以发现，该模型训练很快就得到了较好的模型。

6.7.3 工作原理

使用神经网络训练模型数据有利有弊。神经网络算法模型比先前的算法模型收敛得更快，并在某些场景下更准确，但是同时也要付出代价：我们需要训练更多的模型变量，并且极有可能过拟合。我们从图 6-8 中就可以发现，训练集的准确度在持续地缓慢增加，然而测试集的准确度有时轻微增加，有时会减小。

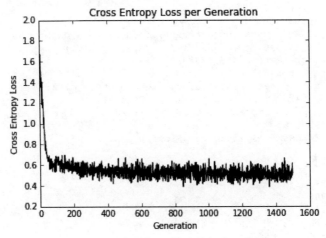

图 6-7　迭代训练 1500 次的损失函数图

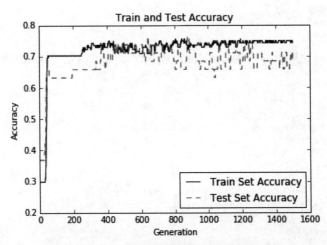

图 6-8　训练集和测试集的准确度图

为了解决欠拟合的问题，我们可以增加训练模型的深度或者迭代训练的次数。为了解决过拟合的问题，我们可以增加更多的数据或者使用正则化技术。

另外，要注意的一点是，神经网络算法模型变量并不像线性模型那样具有可解释性。神经网络算法模型的系数比线性模型更难解释其在算法模型中的特征意义。

6.8　用 TensorFlow 基于神经网络实现井字棋

为了展示如何应用神经网络算法模型，我们将使用神经网络来学习优化井字棋（Tic Tac

Toe）的移动步骤。需要明确的是，井字棋是一种决策性游戏，并且走棋步骤优化是确定的。

6.8.1 开始

为了训练神经网络模型，我们需要一组基于不同棋盘位置的最佳落子点，也就是棋谱。考虑到棋盘的对称性，通过只关心不对称的棋盘位置来简化棋盘。井字棋的非单位变换（考虑几何变换）可以通过 90 度、180 度、270 度、Y 轴对称和 X 轴对称旋转获得。如果这个假设成立，我们使用一系列的棋盘位置列表和对应的最佳落子点，应用两个随机变换，然后赋值给神经网络算法模型学习。

 井字棋是一种决策类游戏，注意，先下者要么赢，要么继续走棋。我们希望能训练一个算法模型给出最佳走棋，使得棋局继续。

在本例中，棋盘走棋一方"×"用"1"表示，对手"O"用"–1"表示，空格棋用"0"表示。图 6-9 展示了棋盘的表示方式和走棋：

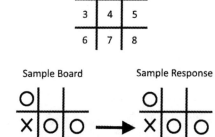

图 6-9　展示棋盘和走棋的表示方式。注意，× = 1，O = –1，空格棋为 0。棋盘位置索引的起始位置标为 0

除了计算模型损失之外，我们将用两种方法来检测算法模型的性能：第一种检测方法是，从训练集中移除一个位置，然后优化走棋。这能看出神经网络算法模型能否生成以前未有过的走棋（即该走棋不在训练集中）；第二种评估的方法是，直接实战井字棋游戏看是否能赢。

不同的棋盘位置列表和对应的最佳落子点数据在 GitHub（https://github.com/nfmcclure/tensorflow_cookbook/tree/master/06_Neural_Networks/08_Learning_Tic_Tac_Toe）和 Packt 代码库（https://github.com/PacktPublishing/TensorFlow-Machine-Learning-Cookbook-Second-Edition）中可以查看。

6.8.2 动手做

1. 导入必要的编程库,代码如下:

```
import tensorflow as tf
import matplotlib.pyplot as plt
import csv
import random
import numpy as np
import random
```

2. 声明训练模型的批量大小,代码如下:

```
batch_size = 50
```

3. 为了让棋盘看起来更清楚,我们创建一个井字棋的打印函数,代码如下:

```
def print_board(board):
    symbols = ['O', ' ', 'X']
    board_plus1 = [int(x) + 1 for x in board]
    board_line1 = ' {} | {} | {}'.format(symbols[board_plus1[0]],
                                          symbols[board_plus1[1]],
                                          symbols[board_plus1[2]])
    board_line2 = ' {} | {} | {}'.format(symbols[board_plus1[3]],
                                          symbols[board_plus1[4]],
                                          symbols[board_plus1[5]])
    board_line3 = ' {} | {} | {}'.format(symbols[board_plus1[6]],
                                          symbols[board_plus1[7]],
                                          symbols[board_plus1[8]])
    print(board_line1)
    print('_____')
    print(board_line2)
    print('_____')
    print(board_line3)
```

4. 创建get_symmetry()函数,返回变换之后的新棋盘和最佳落子点,代码如下:

```
def get_symmetry(board, response, transformation):
    '''
    :param board: list of integers 9 long:
     opposing mark = -1
     friendly mark = 1
     empty space = 0
    :param transformation: one of five transformations on a board:
     rotate180, rotate90, rotate270, flip_v, flip_h
    :return: tuple: (new_board, new_response)
    '''

    if transformation == 'rotate180':
        new_response = 8 - response
        return board[::-1], new_response

    elif transformation == 'rotate90':
        new_response = [6, 3, 0, 7, 4, 1, 8, 5, 2].index(response)
        tuple_board = list(zip(*[board[6:9], board[3:6], board[0:3]]))
        return [value for item in tuple_board for value in item], new_response
```

```
        elif transformation == 'rotate270':
            new_response = [2, 5, 8, 1, 4, 7, 0, 3, 6].index(response)
            tuple_board = list(zip(*[board[0:3], board[3:6],
board[6:9]]))[::-1]
            return [value for item in tuple_board for value in item],
new_response

        elif transformation == 'flip_v':
            new_response = [6, 7, 8, 3, 4, 5, 0, 1, 2].index(response)
            return board[6:9] +  board[3:6] + board[0:3], new_response

        elif transformation == 'flip_h':
        # flip_h = rotate180, then flip_v
            new_response = [2, 1, 0, 5, 4, 3, 8, 7, 6].index(response)
            new_board = board[::-1]
            return new_board[6:9] +  new_board[3:6] + new_board[0:3],
new_response

        else:
            raise ValueError('Method not implmented.')
```

5. 棋盘位置列表和对应的最佳落子点数据位于 .csv 文件中。我们将创建 get_moves_from_csv() 函数来加载文件中的棋盘和最佳落子点数据，并保存成元组，代码如下：

```
def get_moves_from_csv(csv_file):
    '''
    :param csv_file: csv file location containing the boards w/
responses
    :return: moves: list of moves with index of best response
    '''
    moves = []
    with open(csv_file, 'rt') as csvfile:
        reader = csv.reader(csvfile, delimiter=',')
        for row in reader:
            moves.append(([int(x) for x in row[0:9]],int(row[9])))
    return moves
```

6. 创建一个 get_rand_move() 函数，返回一个随机变换棋盘和落子点，代码如下：

```
def get_rand_move(moves, rand_transforms=2):
    # This function performs random transformations on a board.
    (board, response) = random.choice(moves)
    possible_transforms = ['rotate90', 'rotate180', 'rotate270',
'flip_v', 'flip_h']
    for i in range(rand_transforms):
        random_transform = random.choice(possible_transforms)
        (board, response) = get_symmetry(board, response,
random_transform)
    return board, response
```

7. 初始化计算图会话，加载数据文件，创建训练集，代码如下：

```
sess = tf.Session()
moves = get_moves_from_csv('base_tic_tac_toe_moves.csv')
# Create a train set:
train_length = 500
train_set = []
for t in range(train_length):
    train_set.append(get_rand_move(moves))
```

8. 前面提到，我们将从训练集中移除一个棋盘位置和对应的最佳落子点，来看训练的模型是否可以生成最佳走棋。下面棋盘的最佳落子点是棋盘位置索引为 6 的位置，代码如下：

```
test_board = [-1, 0, 0, 1, -1, -1, 0, 0, 1]
train_set = [x for x in train_set if x[0] != test_board]
```

9. 创建 init_weights() 函数和 model() 函数，分别实现初始化模型变量和模型操作。注意，模型中并没有包含 softmax() 激励函数，因为 softmax() 激励函数会在损失函数中出现，代码如下：

```
def init_weights(shape):
    return tf.Variable(tf.random_normal(shape))
def model(X, A1, A2, bias1, bias2):
    layer1 = tf.nn.sigmoid(tf.add(tf.matmul(X, A1), bias1))
    layer2 = tf.add(tf.matmul(layer1, A2), bias2)
    return layer2
```

10. 声明占位符、变量和模型，代码如下：

```
X = tf.placeholder(dtype=tf.float32, shape=[None, 9])
Y = tf.placeholder(dtype=tf.int32, shape=[None])
A1 = init_weights([9, 81])
bias1 = init_weights([81])
A2 = init_weights([81, 9])
bias2 = init_weights([9])
model_output = model(X, A1, A2, bias1, bias2)
```

11. 声明算法模型的损失函数，该函数是最后输出的逻辑变换的平均 softmax 值。然后声明训练步长和优化器。为了将来可以和训练好的模型对局，我们也需要创建预测操作，代码如下：

```
loss = tf.reduce_mean(tf.nn.sparse_softmax_cross_entropy_with_logits(logits=model_output, labels=Y))
train_step = tf.train.GradientDescentOptimizer(0.025).minimize(loss)
prediction = tf.argmax(model_output, 1)
```

12. 初始化变量，遍历迭代训练神经网络模型，代码如下：

```
# Initialize variables
init = tf.global_variables_initializer()
sess.run(init)
loss_vec = []
for i in range(10000):
    # Select random indices for batch
    rand_indices = np.random.choice(range(len(train_set)), batch_size, replace=False)
    # Get batch
    batch_data = [train_set[i] for i in rand_indices]
    x_input = [x[0] for x in batch_data]
    y_target = np.array([y[1] for y in batch_data])
    # Run training step
    sess.run(train_step, feed_dict={X: x_input, Y: y_target})
    # Get training loss
    temp_loss = sess.run(loss, feed_dict={X: x_input, Y: y_target})
    loss_vec.append(temp_loss)
```

```
if i%500==0:
    print('iteration ' + str(i) + ' Loss: ' + str(temp_loss))
```

13. 绘制模型训练的损失函数，代码如下（对应的图见图6-10）：

```
plt.plot(loss_vec, 'k-', label='Loss')
plt.title('Loss (MSE) per Generation')
plt.xlabel('Generation')
plt.ylabel('Loss')
plt.show()
```

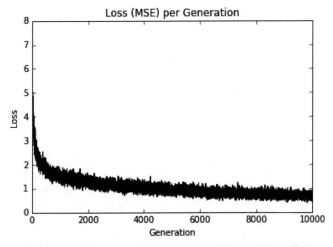

图 6-10　迭代 10 000 次训练的井字棋模型的损失函数图

14. 为了测试模型，将展示如何在测试棋盘（从训练集中移除的数据）使用。我们希望看到模型能生成预测落子点的索引，并且索引值为 6。在大部分情况下，模型都会成功预测，代码如下：

```
test_boards = [test_board]
feed_dict = {X: test_boards}
logits = sess.run(model_output, feed_dict=feed_dict)
predictions = sess.run(prediction, feed_dict=feed_dict)
print(predictions)
```

15. 输出结果如下：

[6]

16. 为了能够评估训练模型，我们计划和训练好的模型进行对局。为了实现该功能，我们创建一个函数来检测是否赢了棋局，这样程序才能在该结束的时间喊停，代码如下：

```
def check(board):
    wins = [[0,1,2], [3,4,5], [6,7,8], [0,3,6], [1,4,7], [2,5,8],
    [0,4,8], [2,4,6]]
    for i in range(len(wins)):
        if board[wins[i][0]]==board[wins[i][1]]==board[wins[i][2]]==1.:
            return 1
```

```
        elif
board[wins[i][0]]==board[wins[i][1]]==board[wins[i][2]]==-1.:
            return 1
    return 0
```

17. 现在遍历迭代，同训练模型进行对局。起始棋盘为空棋盘，即为全 0 值；然后询问棋手要在哪个位置落棋子，即输入 0-8 的索引值；接着将其传入训练模型进行预测。对于模型的走棋，我们获得了多个可能的预测。最后显示井字棋游戏的样例。对于该游戏来说，我们发现训练的模型表现得并不理想，代码如下：

```
game_tracker = [0., 0., 0., 0., 0., 0., 0., 0., 0.]
win_logical = False
num_moves = 0
while not win_logical:
    player_index = input('Input index of your move (0-8): ')
    num_moves += 1
    # Add player move to game
    game_tracker[int(player_index)] = 1.
    # Get model's move by first getting all the logits for each index
    [potential_moves] = sess.run(model_output, feed_dict={X: [game_tracker]})
    # Now find allowed moves (where game tracker values = 0.0)
    allowed_moves = [ix for ix,x in enumerate(game_tracker) if x==0.0]
    # Find best move by taking argmax of logits if they are in allowed moves
    model_move = np.argmax([x if ix in allowed_moves else -999.0 for ix,x in enumerate(potential_moves)])
    # Add model move to game
    game_tracker[int(model_move)] = -1.
    print('Model has moved')
    print_board(game_tracker)
    # Now check for win or too many moves
    if check(game_tracker)==1 or num_moves>=5:
        print('Game Over!')
        win_logical = True
```

18. 人机交互的输出结果如下：

```
Input index of your move (0-8): 4
Model has moved
 O |   |
_____
   | X |
_____
   |   |
Input index of your move (0-8): 6
Model has moved
 O |   |
_____
   | X |
_____
 X |   | O
Input index of your move (0-8): 2
```

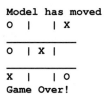

6.8.3 工作原理

我们训练一个神经网络模型来玩井字棋游戏，该模型需要传入棋盘位置，其中棋盘的位置是用一个九维向量来表示的。然后预测最佳落子点。我们需要赋值可能的井字棋棋盘，应用随机转换来增加训练集的大小。

为了测试算法模型，我们移除一个棋盘位置列表和对应的最佳落子点，然后观察训练模型能否生成预测的最佳落棋点。最后，我们也和训练模型进行对比，但是结果并不理想，我们仍然需要尝试不同的架构和训练方法来提高效果。

CHAPTER 7

第 7 章

自然语言处理

本章将介绍 TensorFlow 在文本处理中的使用。先介绍利用词袋方法进行词嵌入的工作原理，然后讲解更高级的词嵌入方法，比如，Word2Vec 和 Doc2Vec。

本章主要学习以下知识：
- 词袋的使用
- TensorFlow 实现 TF-IDF 算法
- TensorFlow 实现 skip-gram 模型
- TensorFlow 实现 CBOW 词嵌入模型
- TensorFlow 实现 Word2Vec 预测
- TensorFlow 实现基于 Doc2Vec 的情感分析

注意，读者可以在 Github 上找到本章的所有代码，网址：https://github.com/nfmcclure/tensorflw-cookbook 或 https://github.com/ PacktPublishing/ TensorFlow- Machine- Learning- Cookbook- Second- Edition。

7.1 简介

到目前为止，我们介绍的机器学习算法都是在数值型输入上进行操作。如果我们想进行文本处理，就必须找到一种方法将文本转化成数字。有许多方法可以实现该功能，这里我们将介绍一些常用的方法。

假设要处理语句 TensorFlow makes machine learning easy，我们可以按观察到的顺序将单词转换成数字，假设转换成 1 2 3 4 5。然后当我们看到一个新的句子 *machine learning is easy*，那么很容易将其翻译成 3 4 0 5，用 0 代表不能识别的单词。在前面的两个例子中，我们把词汇限制为 6 个数字。对于长文本，我们可以选择保留多少单词，一般选择的单词是使用频率高的词汇，并用 0 标注其他任意单词。

如果单词 learning 用数字 4 表示，单词 makes 用数字 2 表示，那很自然地认为单词 learning 是单词 makes 的两倍。因为我们不期望两个单词之间有数值关系，所以我们假设这些数字仅仅代表分类，而不具有数值关系。

另一个问题是，两句话的长度不一样。我们的算法模型期望的输入是相同长度的语句，为了解决这个问题，我们把句子转成一个稀疏向量。该稀疏向量的规则是，如果对应索引上的单词存在，则对应的索引位置的值为 1。

TensorFlow	makes	machine	learning	easy
1	2	3	4	5

`first_sentence = [0,1,1,1,1,1]`

为了进一步解释上述向量，我们的词汇由 6 个不同的单词组成（5 个已知，1 个未知）。每个单词赋值 0 或 1，其中 0 代表句子中没有出现该单词，而 1 代表该单词至少出现 1 次。所以 0 意味着单词不出现，1 意味着出现。

Machine	learning	is	easy
3	4	0	5

`second_sentence = [1,0,0,1,1,1]`

这种方法的缺点是损失了语句中单词顺序的特征。TensorFlow makes machine learning easy 和 machine learning makes TensorFlow easy，不同的两句话却具有相同句子稀疏向量。

值得注意的是，这些向量的长度是相等的，并且与我们所选的词汇量一致。一般情况下，我们会选择非常大的词汇量，所以这些句子向量非常稀疏。这种词嵌入方法称为"词袋"，下一节将会介绍。

另外一个缺点是，单词 is 和 TensorFlow 具有相同的数值化索引值 1，但是我们很明显地看到单词 is 要比单词 TensorFlow 的重要性弱。

我们将在本章中阐述不同的词嵌入方法，试图解决上面的问题。下面先来看下"词袋"的实现。

7.2 词袋的使用

本节将展示如何使用 TensorFlow 的"词袋"嵌入。"词袋"嵌入映射的原理在上一节已经讲过，这里只展示如何使用"词袋"嵌入来进行垃圾邮件的预测。

7.2.1 开始

为了阐述在文本数据集上如何使用"词袋"，我们将使用 UCI 机器学习数据库中的垃圾邮件文本数据集（https://archive.ics.uci.edu/ml/datasets/SWS+Spam+Collection），该垃圾邮件数据集中有正常邮件和垃圾邮件。我们将下载该数据集，存储好以备后用。然后用"词袋"的方法预测一条文本是否为垃圾邮件。我们将使用不含隐藏层的逻辑模型来训练"词袋"，采用批量大小为 1 的随机训练，并计算留存的测试集的准确度。

7.2.2 动手做

对于本例，首先获取数据集，归一化和分割文本数据，运行词嵌入函数，训练逻辑函数来预测垃圾邮件。

1. 导入必要的编程库。本例中需要 .zip file 库来解压从 UCI 机器学习数据库中下载的 .zip 文件，代码如下：

```
import tensorflow as tf
import matplotlib.pyplot as plt
import os
import numpy as np
import csv
import string
import requests
import io
from zipfile import ZipFile
from tensorflow.contrib import learn
sess = tf.Session()
```

2. 为了让脚本运行时不用每次都去下载文本数据，我们将下载文件并存储，并检查之前是否保存过。该步骤避免了文本数据的重复下载。下载完文本数据集后，抽取输入数据和目标数据，并调整目标值（垃圾邮件（spam）置为 1，正常邮件（ham）置为 0）。具体代码如下：

```
save_file_name = os.path.join('temp','temp_spam_data.csv')
if os.path.isfile(save_file_name):
 text_data = []
    with open(save_file_name, 'r') as temp_output_file:
        reader = csv.reader(temp_output_file)
        for row in reader:
            text_data.append(row)
else:
    zip_url = 'http://archive.ics.uci.edu/ml/machine-learning-databases/00228/smsspamcollection.zip'
    r = requests.get(zip_url)
    z = ZipFile(io.BytesIO(r.content))
    file = z.read('SMSSpamCollection')
    # Format Data
    text_data = file.decode()
    text_data = text_data.encode('ascii',errors='ignore')
    text_data = text_data.decode().split('\n')
    text_data = [x.split('\t') for x in text_data if len(x)>=1]
    # And write to csv
    with open(save_file_name, 'w') as temp_output_file:
        writer = csv.writer(temp_output_file)
        writer.writerows(text_data)
texts = [x[1] for x in text_data]
target = [x[0] for x in text_data]
# Relabel 'spam' as 1, 'ham' as 0
target = [1 if x=='spam' else 0 for x in target]
```

3. 为了减小词汇量大小，我们对文本进行归一化处理。移除文本中大小写和数字的影响，代码如下：

```
# Convert to lower case
texts = [x.lower() for x in texts]
# Remove punctuation
```

```
texts = [''.join(c for c in x if c not in string.punctuation) for x
in texts]
# Remove numbers
texts = [''.join(c for c in x if c not in '0123456789') for x in
texts]
# Trim extra whitespace
texts = [' '.join(x.split()) for x in texts]
```

4. 计算最长句子大小。我们使用文本数据集的文本长度直方图（见图7-1），并取最佳截止点（本例中取值为25个单词），代码如下：

```
# Plot histogram of text lengths
text_lengths = [len(x.split()) for x in texts]
text_lengths = [x for x in text_lengths if x < 50]
plt.hist(text_lengths, bins=25)
plt.title('Histogram of # of Words in Texts')
sentence_size = 25
min_word_freq = 3
```

图7-1　文本数据中的单词数的直方图

> 我们用该直方图选出最大单词长度。本例设为25个单词，但是也可以设为30或者40。

5. TensorFlow 内置转换函数 VocabularyProcessor()，该函数位于 learn.preprocessing 库，注意，使用此函数时可能会出现弃用警告。代码如下：

```
vocab_processor =
learn.preprocessing.VocabularyProcessor(sentence_size,
min_frequency=min_word_freq)
vocab_processor.fit_transform(texts)
transformed_texts = np.array([x for x in
vocab_processor.transform(texts)])
embedding_size = len(np.unique(transformed_texts))
```

6. 分割数据集为训练集（80%）和测试集（20%），代码如下：

```
train_indices = np.random.choice(len(texts), round(len(texts)*0.8),
replace=False)
```

```
test_indices = np.array(list(set(range(len(texts))) -
set(train_indices)))
texts_train = [x for ix, x in enumerate(texts) if ix in
train_indices]
texts_test = [x for ix, x in enumerate(texts) if ix in
test_indices]
target_train = [x for ix, x in enumerate(target) if ix in
train_indices]
target_test = [x for ix, x in enumerate(target) if ix in
test_indices]
```

7. 声明词嵌入矩阵。将句子单词转成索引,再将索引转成 one-hot 向量,该向量为单位矩阵所创造。我们使用该矩阵为每个单词查找稀疏向量,并加入到词稀疏向量,代码如下:

```
identity_mat = tf.diag(tf.ones(shape=[embedding_size]))
```

8. 因为最后要进行逻辑回归预测垃圾邮件的概率,所以我们需要声明逻辑回归变量。然后声明占位符,注意 x_data 输入占位符是整数类型,因为它被用来查找单位矩阵的行索引,而 TensorFlow 要求其为整数类型,代码如下:

```
A = tf.Variable(tf.random_normal(shape=[embedding_size,1]))
b = tf.Variable(tf.random_normal(shape=[1,1]))
# Initialize placeholders
x_data = tf.placeholder(shape=[sentence_size], dtype=tf.int32)
y_target = tf.placeholder(shape=[1, 1], dtype=tf.float32)
```

9. 使用 TensorFlow 的嵌入查找函数来映射句子中的单词索引为单位矩阵的 one-hot 向量。然后把前面的词向量求和,得到句子向量,代码如下:

```
x_embed = tf.nn.embedding_lookup(identity_mat, x_data)
x_col_sums = tf.reduce_sum(x_embed, 0)
```

10. 有了每个句子的固定长度的句子向量之后,我们进行逻辑回归训练。为此,声明逻辑回归算法模型。因为一次做一个数据点的随机训练,所以扩展输入数据的维度,并进行线性回归操作。记住,TensorFlow 中的损失函数已经包含了 sigmoid 激励函数,所以我们不需要在输出时加入激励函数,代码如下:

```
x_col_sums_2D = tf.expand_dims(x_col_sums, 0)
model_output = tf.add(tf.matmul(x_col_sums_2D, A), b)
```

11. 声明训练模型的损失函数、预测函数和优化器,代码如下:

```
loss =
tf.reduce_mean(tf.nn.sigmoid_cross_entropy_with_logits(logits=model
_output, labels=y_target))
# Prediction operation
prediction = tf.sigmoid(model_output)
# Declare optimizer
my_opt = tf.train.GradientDescentOptimizer(0.001)
train_step = my_opt.minimize(loss)
```

12. 接下来初始化计算图中的变量,代码如下:

```
init = tf.global_variables_initializer()
sess.run(init)
```

13. 开始迭代训练。TensorFlow 的内建函数 vocab_processor.fit() 是一个符合本例的生成器。我们将使用该函数来进行随机训练逻辑回归模型。为了得到准确度的趋势，我们保留最近 50 次迭代的平均值。如果只绘制当前值，我们会依赖预测训练数据点是否正确而得到 1 或者 0 的值，代码如下：

```
loss_vec = []
train_acc_all = []
train_acc_avg = []
for ix, t in enumerate(vocab_processor.fit_transform(texts_train)):
    y_data = [[target_train[ix]]]
    sess.run(train_step, feed_dict={x_data: t, y_target: y_data})
    temp_loss = sess.run(loss, feed_dict={x_data: t, y_target: y_data})
    loss_vec.append(temp_loss)
    if (ix+1)%10==0:
        print('Training Observation #{}: Loss= {}'.format(ix+1, temp_loss))
    # Keep trailing average of past 50 observations accuracy
    # Get prediction of single observation
    [[temp_pred]] = sess.run(prediction, feed_dict={x_data:t, y_target:y_data})
    # Get True/False if prediction is accurate
    train_acc_temp = target_train[ix]==np.round(temp_pred)
    train_acc_all.append(train_acc_temp)
    if len(train_acc_all) >= 50:
        train_acc_avg.append(np.mean(train_acc_all[-50:]))
```

14. 训练结果如下：

```
Starting Training Over 4459 Sentences.
Training Observation #10: Loss = 5.45322
Training Observation #20: Loss = 3.58226
Training Observation #30: Loss = 0.0
...
Training Observation #4430: Loss = 1.84636
Training Observation #4440: Loss = 1.46626e-05
Training Observation #4450: Loss = 0.045941
```

15. 为了得到测试集的准确度，我们重复处理过程，对测试文本只进行预测操作，而不进行训练操作，代码如下：

```
print('Getting Test Set Accuracy')
test_acc_all = []
for ix, t in enumerate(vocab_processor.fit_transform(texts_test)):
    y_data = [[target_test[ix]]]
    if (ix+1)%50==0:
        print('Test Observation #{}'.format(ix+1))
    # Keep trailing average of past 50 observations accuracy
    # Get prediction of single observation
    [[temp_pred]] = sess.run(prediction, feed_dict={x_data:t, y_target:y_data})
    # Get True/False if prediction is accurate
    test_acc_temp = target_test[ix]==np.round(temp_pred)
    test_acc_all.append(test_acc_temp)
print('\nOverall Test Accuracy: {}'.format(np.mean(test_acc_all)))
```

```
Getting Test Set Accuracy For 1115 Sentences.
Test Observation #10
Test Observation #20
Test Observation #30
...
Test Observation #1000
Test Observation #1050
Test Observation #1100
Overall Test Accuracy: 0.8035874439461883
```

7.2.3 工作原理

在本例中,我们处理来自 UCI 机器学习库中的垃圾邮件文本数据。使用 TensorFlow 的词汇处理函数来创建标准的词汇和句子向量,该句子向量是文本单词向量的总和。我们在逻辑回归算法模型中进行句子向量的训练,预测垃圾邮件获得了大约 80% 的准确度。

7.2.4 延伸学习

值得注意的是本例中限制句子(或文本)长度的动机。本例限制文本大小为 25 个单词。这是"词袋"的常用实践,因为文本长度影响到预测结果。想象一下,如果我们发现一个单词(如 meeting)预测为正常邮件,然后该单词出现许多次,那就变成垃圾邮件了。

事实上,这在数据不均衡的情况下最普遍。在数据不均衡的情况下,垃圾邮件变得很难找到,而正常邮件很容易发现。由于这个事实,我们创建的词汇就会严重倾斜到正常邮件数据这边。如果我们不限制文本长度,那发垃圾邮件者会利用这一点,创建很长的文本,而长文本触发逻辑回归模型中非垃圾邮件因素的概率更高。

下一节试图以词频来更好地解决该问题。

7.3 用 TensorFlow 实现 TF-IDF 算法

因为我们为每个单词选择词嵌入,所以可能需要调整某些特定单词的权重。在这种策略下,应提高有用词汇的权重,降低常用单词或者无意义单词的权重。本节将展示使用该方法的词嵌入。

7.3.1 开始

TF-IDF(Text Frequency-Inverse Document Frequency)算法表示为词频和逆文档频率的乘积。

上一节介绍了"词袋"的方法,句子中每个单词出现一次就分配一个值"1"。这可能不是太理想的方法,每类句子(前面的例子中分垃圾邮件和非垃圾邮件)可能拥有相同频率的 the、and 和其他的单词,而 Viagra 和 sale 这类词可能在判断邮件是否为垃圾邮件时应该增加权重。

首先，我们考虑词频（TF）。词频是某个单词在文档中出现的频率，其目的是找到每个单词的权重。

但是 the 和 and 之类的单词在每个文档中出现的频率都很高，我们需要降低这些单词的权重。所以，用词频（TF）乘以逆文档频率（其为所有包含该单词的文档频率的倒数）即可找到重要的单词。但是因为语料库往往数量巨大，普遍的做法是对文档频率求 log。这样就得到每个单词在每个文档中的 TF-IDF 公式：

$$w_{tf-idf} = w_{tf} \cdot \log\left(\frac{1}{w_{df}}\right)$$

其中，W_{tf} 是文档的词频，W_{df} 是包含该单词的所有文档的总频率。TF-IDF 值高代表着单词在文档中的重要性。

创建 TF-IDF 向量要求向内存加载所有的文本数据，并在开始训练模型之前计算每个单词出现的次数。因此不能完全依赖 TensorFlow，我们将使用 scikit-learn 创建 TF-IDF 向量，但是采用 TensorFlow 拟合逻辑回归模型。

7.3.2 动手做

1. 导入必要的编程库。本例中会导入 scikit-learn 的 TF-IDF 处理模块处理文本数据集，代码如下：

```
import tensorflow as tf
import matplotlib.pyplot as plt
import csv
import numpy as np
import os
import string
import requests
import io
import nltk
from zipfile import ZipFile
from sklearn.feature_extraction.text import TfidfVectorizer
```

2. 创建一个计算图会话，声明批量大小和词汇的最大长度，代码如下：

```
sess = tf.Session()
batch_size= 200
max_features = 1000
```

3. 加载文本数据集。可以从网站下载或者从上次保存的 temp 文件夹加载，代码如下：

```
save_file_name = os.path.join('temp','temp_spam_data.csv')
if os.path.isfile(save_file_name):
    text_data = []
    with open(save_file_name, 'r') as temp_output_file:
        reader = csv.reader(temp_output_file)
        for row in reader:
            text_data.append(row)
else:
    zip_url =
'http://archive.ics.uci.edu/ml/machine-learning-databases/00228/sms
```

```
spamcollection.zip'
    r = requests.get(zip_url)
    z = ZipFile(io.BytesIO(r.content))
    file = z.read('SMSSpamCollection')
    # Format Data
    text_data = file.decode()
    text_data = text_data.encode('ascii',errors='ignore')
    text_data = text_data.decode().split('\n')
    text_data = [x.split('\t') for x in text_data if len(x)>=1]
    # And write to csv
    with open(save_file_name, 'w') as temp_output_file:
        writer = csv.writer(temp_output_file)
        writer.writerows(text_data)
texts = [x[1] for x in text_data]
target = [x[0] for x in text_data]
# Relabel 'spam' as 1, 'ham' as 0
target = [1. if x=='spam' else 0. for x in target]
```

4. 像之前一样，通过将所有字符转成小写，剔除标点符号和数字，以此减小词汇长度，代码如下：

```
# Lower case
texts = [x.lower() for x in texts]
# Remove punctuation
texts = [''.join(c for c in x if c not in string.punctuation) for x in texts]
# Remove numbers
texts = [''.join(c for c in x if c not in '0123456789') for x in texts]
# Trim extra whitespace
texts = [' '.join(x.split()) for x in texts]
```

5. 为了使用 scikit-learn 的 TF-IDF 处理函数，我们需要将句子进行分割（即将句子切分为相应的单词）。nltk 包可以提供非常棒的分词器来实现分词功能，代码如下：

```
def tokenizer(text):
    words = nltk.word_tokenize(text)
    return words
# Create TF-IDF of texts
tfidf = TfidfVectorizer(tokenizer=tokenizer, stop_words='english', max_features=max_features)
sparse_tfidf_texts = tfidf.fit_transform(texts)
```

6. 分割数据集为训练集和测试集，代码如下：

```
train_indices = np.random.choice(sparse_tfidf_texts.shape[0], round(0.8*sparse_tfidf_texts.shape[0]), replace=False)
test_indices =
np.array(list(set(range(sparse_tfidf_texts.shape[0])) - set(train_indices)))
texts_train = sparse_tfidf_texts[train_indices]
texts_test = sparse_tfidf_texts[test_indices]
target_train = np.array([x for ix, x in enumerate(target) if ix in train_indices])
target_test = np.array([x for ix, x in enumerate(target) if ix in test_indices])
```

7. 声明逻辑回归模型的变量和数据集的占位符，代码如下：

```
A = tf.Variable(tf.random_normal(shape=[max_features,1]))
b = tf.Variable(tf.random_normal(shape=[1,1]))
# Initialize placeholders
x_data = tf.placeholder(shape=[None, max_features], dtype=tf.float32)
y_target = tf.placeholder(shape=[None, 1], dtype=tf.float32)
```

8. 声明算法模型操作和损失函数。注意，逻辑回归算法的sigmoid部分是在损失函数中实现的，代码如下：

```
model_output = tf.add(tf.matmul(x_data, A), b)
loss = tf.reduce_mean(tf.nn.sigmoid_cross_entropy_with_logits(logits=model_output, labels=y_target))
```

9. 为计算图增加预测函数和准确度函数（可以让我们看到模型训练过程中训练集和测试集的准确度），代码如下：

```
prediction = tf.round(tf.sigmoid(model_output))
predictions_correct = tf.cast(tf.equal(prediction, y_target), tf.float32)
accuracy = tf.reduce_mean(predictions_correct)
```

10. 声明优化器，初始化计算图中的变量，代码如下：

```
my_opt = tf.train.GradientDescentOptimizer(0.0025)
train_step = my_opt.minimize(loss)
# Intitialize Variables
init = tf.global_variables_initializer()
sess.run(init)
```

11. 遍历迭代训练模型10 000次，记录测试集和训练集损失，以及每迭代100次的准确度，然后每迭代500次就打印状态信息，代码如下：

```
train_loss = []
test_loss = []
train_acc = []
test_acc = []
i_data = []
for i in range(10000):
    rand_index = np.random.choice(texts_train.shape[0], size=batch_size)
    rand_x = texts_train[rand_index].todense()
    rand_y = np.transpose([target_train[rand_index]])
    sess.run(train_step, feed_dict={x_data: rand_x, y_target: rand_y})
    # Only record loss and accuracy every 100 generations
    if (i+1)%100==0:
        i_data.append(i+1)
        train_loss_temp = sess.run(loss, feed_dict={x_data: rand_x, y_target: rand_y})
        train_loss.append(train_loss_temp)
        test_loss_temp = sess.run(loss, feed_dict={x_data: texts_test.todense(), y_target: np.transpose([target_test])})
        test_loss.append(test_loss_temp)
```

```
        train_acc_temp = sess.run(accuracy, feed_dict={x_data:
rand_x, y_target: rand_y})
        train_acc.append(train_acc_temp)
        test_acc_temp = sess.run(accuracy, feed_dict={x_data:
texts_test.todense(), y_target: np.transpose([target_test])})
        test_acc.append(test_acc_temp)
    if (i+1)%500==0:
        acc_and_loss = [i+1, train_loss_temp, test_loss_temp,
train_acc_temp, test_acc_temp]
        acc_and_loss = [np.round(x,2) for x in acc_and_loss]
        print('Generation # {}. Train Loss (Test Loss): {:.2f}
({:.2f}). Train Acc (Test Acc): {:.2f}
({:.2f})'.format(*acc_and_loss))
```

12. 输出结果如下：

```
Generation # 500. Train Loss (Test Loss): 0.69 (0.73). Train Acc
(Test Acc): 0.62 (0.57)
Generation # 1000. Train Loss (Test Loss): 0.62 (0.63). Train Acc
(Test Acc): 0.68 (0.66)
...
Generation # 9500. Train Loss (Test Loss): 0.39 (0.45). Train Acc
(Test Acc): 0.89 (0.85)
Generation # 10000. Train Loss (Test Loss): 0.48 (0.45). Train Acc
(Test Acc): 0.84 (0.85)
```

绘制训练集和测试集的准确度和损失函数（见图 7-2 和图 7-3）：

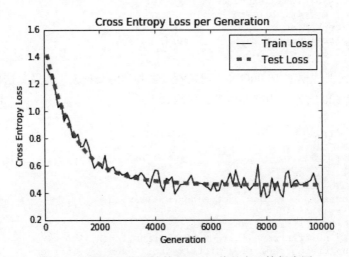

图 7-2　逻辑回归模型计算 TF-IDF 值的交叉熵损失图

7.3.3　工作原理

使用 TF-IDF 值来进行模型训练，其预测函数的准确度由上一节"词袋"方法的 80% 提高到 90%。改善的原因正是我们采用 scikit-learn 的 TF-IDF 处理函数并用 TF-IDF 值进行逻辑回归模型训练。

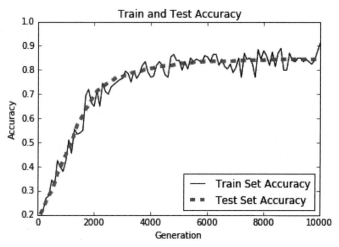

图 7-3 基于 TF-IDF 值的垃圾邮件逻辑回归模型的训练集和测试集准确度图

7.3.4 延伸学习

本节解决了单词重要性的问题,但是没有解决单词顺序的问题。"词袋"方法和 TF-IDF 方法都未考虑句子中单词的顺序这一特征。下一节引入的 Word2Vec 技术试图解决该问题。

7.4 用 TensorFlow 实现 skip-gram 模型

在上一节中,我们已经描述了训练模型前的词嵌入。使用神经网络算法可以得到训练过程中词嵌入向量。我们将研究的第一个此类方法是 skip-gram 嵌套。

7.4.1 开始

在前一节中,我们在创建词向量时未考虑相关单词顺序特征。在 2013 年早期,就职于 Google 的 Tomas Mikolov 和另一些研究者发表了一篇论文(https://arxiv.org/abs/1301.3781),并创建了一种词向量方法来解决该问题,该方法被命名为 Word2Vec。

基本的思路是创建词向量来表现单词上下文关系。我们寻找如何理解互相关联的单词。下面是单词如何嵌入的例子:

king – man + woman = queen

India pale ale – hops + malt = stout

如果仅仅考虑每个单词位置之间的关系,那么我们可以得到数值化的表示。如果分析大量连续文档,我们可以看到,单词 king、man 和 queen 是紧密联系的。如果已知 man 和 woman 语义相关联,那我们可以得出单词 man 和 king 的关系就如同单词 woman 和 queen 的关系。

为了找到这种词嵌入,我们使用神经网络算法预测输入单词的上下文单词。我们也能轻松地根据上下文单词预测目标单词,这两种方法都是 Word2Vec 的变体。我们先讲解第一种方法。第一种根据目标单词进行上下文相关单词的预测,称为 skip-gram 模型(见图 7-4)。在下一节中,我们将实现另外一种方法,从上下文相关单词集合中预测目标单词,称为"连续词袋"(continuous bag of words,CBOW)方法:

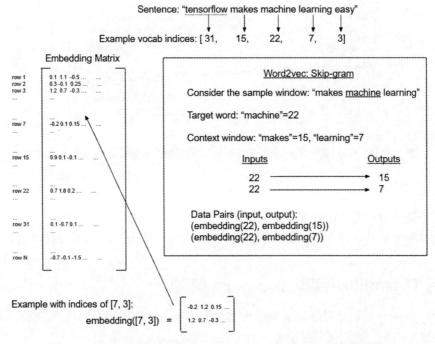

图 7-4　Word2Vec 中的 skip-gram 模型。skip-gram 模型从目标单词中预测上下文窗口大小(两边宽度均为 1)内的单词

本节将使用康奈尔大学的电影影评数据集(http://www.cs.cornell.edu/people/pabo/movie-review-data/)来实现 skip-gram 模型。CBOW 方法将在后面实现。

7.4.2　动手做

本节将创建几个辅助函数:加载数据函数、归一化文本函数、生成词汇表函数和生成批量数据函数。在实现这些函数之后,我们开始训练词向量。我们不预测任何目标变量,而是拟合词向量。

1.导入必要的编程库,开始一个计算图会话,代码如下:

```
import tensorflow as tf
import matplotlib.pyplot as plt
import numpy as np
import random
import os
import string
```

```
import requests
import collections
import io
import tarfile
import urllib.request
from nltk.corpus import stopwords
sess = tf.Session()
```

2. 声明一些模型参数。一次将查找 50 对词嵌入（批量大小）。每个单词的嵌套大小是一个长度为 200 的向量，并且仅仅考虑 10 000 个频率最高的单词（其他单词分类为"unknown"）。我们将迭代训练 50 000 次，每迭代 500 次打印损失函数。然后声明一个 num_sampled 变量，该变量将在损失函数中使用，具体解释见下文。我们也需要声明 skip-gram 模型上下文窗口大小，本例中窗口大小设为 2，即将查找目标单词两边各两个上下文单词。我们使用 Python 的 nltk 包做"停词"步骤。为了检测词向量的性能，选择一些常用的电影影评单词，每迭代 2000 次训练打印最近邻域单词。具体代码如下：

```
batch_size = 50
embedding_size = 200
vocabulary_size = 10000
generations = 50000
print_loss_every = 500
num_sampled = int(batch_size/2)
window_size = 2
stops = stopwords.words('english')
print_valid_every = 2000
valid_words = ['cliche', 'love', 'hate', 'silly', 'sad']
```

3. 声明数据加载函数，该函数会在下载数据前先检测是否已下载过该数据集，如果已下载过，将直接从磁盘加载数据，代码如下：

```
def load_movie_data():
    save_folder_name = 'temp'
    pos_file = os.path.join(save_folder_name, 'rt-polarity.pos')
    neg_file = os.path.join(save_folder_name, 'rt-polarity.neg')
    # Check if files are already downloaded
    if os.path.exists(save_folder_name):
        pos_data = []
        with open(pos_file, 'r') as temp_pos_file:
            for row in temp_pos_file:
                pos_data.append(row)
        neg_data = []
        with open(neg_file, 'r') as temp_neg_file:
            for row in temp_neg_file:
                neg_data.append(row)
    else: # If not downloaded, download and save
        movie_data_url = 'http://www.cs.cornell.edu/people/pabo/movie-review-data/rt-polaritydata.tar.gz'
        stream_data = urllib.request.urlopen(movie_data_url)
        tmp = io.BytesIO()
        while True:
            s = stream_data.read(16384)
            if not s:
                break
```

```
                break
            tmp.write(s)
            stream_data.close()
            tmp.seek(0)
        tar_file = tarfile.open(fileobj=tmp, mode='r:gz')
        pos = tar_file.extractfile('rt-polaritydata/rt-polarity.pos')
        neg = tar_file.extractfile('rt-polaritydata/rt-polarity.neg')
        # Save pos/neg reviews
        pos_data = []
        for line in pos:
pos_data.append(line.decode('ISO-8859-1').encode('ascii',errors='ignore').decode())
        neg_data = []
        for line in neg:
neg_data.append(line.decode('ISO-8859-1').encode('ascii',errors='ignore').decode())
        tar_file.close()
        # Write to file
        if not os.path.exists(save_folder_name):
            os.makedirs(save_folder_name)
        # Save files
        with open(pos_file, 'w') as pos_file_handler:
            pos_file_handler.write(''.join(pos_data))
        with open(neg_file, 'w') as neg_file_handler:
            neg_file_handler.write(''.join(neg_data))
    texts = pos_data + neg_data
    target = [1]*len(pos_data) + [0]*len(neg_data)
    return(texts, target)
texts, target = load_movie_data()
```

4. 创建归一化文本函数。该函数输入一列字符串，转换成小写字符，移除标点符号，移除数字，去除多余的空白字符，并移除"停词"，代码如下：

```
def normalize_text(texts, stops):
    # Lower case
    texts = [x.lower() for x in texts]
    # Remove punctuation
    texts = [''.join(c for c in x if c not in string.punctuation) for x in texts]
    # Remove numbers
    texts = [''.join(c for c in x if c not in '0123456789') for x in texts]
    # Remove stopwords
    texts = [' '.join([word for word in x.split() if word not in (stops)]) for x in texts]
    # Trim extra whitespace
    texts = [' '.join(x.split()) for x in texts]
    return(texts)
texts = normalize_text(texts, stops)
```

5. 为了确保所有电影影评的有效性，我们将检查其中的影评长度。可以强制影评长度为三个单词或者更长长度的单词，代码如下：

```
target = [target[ix] for ix, x in enumerate(texts) if
```

```
    len(x.split()) > 2]
texts = [x for x in texts if len(x.split()) > 2]
```

6. 构建词汇表，创建函数来建立一个单词字典（该单词词典是单词和单词数对）。词频不够的单词（即标记为 unknown 的单词）标记为 RARE，代码如下：

```
def build_dictionary(sentences, vocabulary_size):
    # Turn sentences (list of strings) into lists of words
    split_sentences = [s.split() for s in sentences]
    words = [x for sublist in split_sentences for x in sublist]
    # Initialize list of [word, word_count] for each word, starting with unknown
    count = [['RARE', -1]]
    # Now add most frequent words, limited to the N-most frequent (N=vocabulary size)
    count.extend(collections.Counter(words).most_common(vocabulary_size -1))
    # Now create the dictionary
    word_dict = {}
    # For each word, that we want in the dictionary, add it, then make it the value of the prior dictionary length
    for word, word_count in count:
        word_dict[word] = len(word_dict)
    return(word_dict)
```

7. 创建一个函数将一系列的句子转化成单词索引列表，并将单词索引列表传入嵌入寻找函数，代码如下：

```
def text_to_numbers(sentences, word_dict):
    # Initialize the returned data
    data = []
    for sentence in sentences:
        sentence_data = []
        # For each word, either use selected index or rare word index
        for word in sentence:
            if word in word_dict:
                word_ix = word_dict[word]
            else:
                word_ix = 0
            sentence_data.append(word_ix)
        data.append(sentence_data)
    return data
```

8. 创建单词字典，转换句子列表为单词索引列表，代码如下：

```
word_dictionary = build_dictionary(texts, vocabulary_size)
word_dictionary_rev = dict(zip(word_dictionary.values(), word_dictionary.keys()))
text_data = text_to_numbers(texts, word_dictionary)
```

9. 从预处理的单词词典中，查找第二步中选择的验证单词的索引，代码如下：

```
valid_examples = [word_dictionary[x] for x in valid_words]
```

10. 创建函数返回 skip-gram 模型的批量数据。我们希望在单词对上训练模型，该单词对中第一个单词为训练输入（即窗口中央的目标单词）；另一个单词为窗口中所选的单词。

例如，句子"the cat in the hat"，如果 in 为目标单词，那么该例句的上下文窗口大小为 2 的单词对（输入单词，输出单词）为：(the, in)、(cat, in)、(the, in)、(hat, in)。具体代码如下：

```
def generate_batch_data(sentences, batch_size, window_size,
method='skip_gram'):
    # Fill up data batch
    batch_data = []
    label_data = []
    while len(batch_data) < batch_size:
        # select random sentence to start
        rand_sentence = np.random.choice(sentences)
        # Generate consecutive windows to look at
        window_sequences = [rand_sentence[max((ix-
window_size),0):(ix+window_size+1)] for ix, x in
enumerate(rand_sentence)]
        # Denote which element of each window is the center word of
interest
        label_indices = [ix if ix<window_size else window_size for
ix,x in enumerate(window_sequences)]
        # Pull out center word of interest for each window and
create a tuple for each window
        if method=='skip_gram':
            batch_and_labels = [(x[y], x[:y] + x[(y+1):]) for x,y
in zip(window_sequences, label_indices)]
            # Make it in to a big list of tuples (target word,
surrounding word)
            tuple_data = [(x, y_) for x,y in batch_and_labels for
y_ in y]
        else:
            raise ValueError('Method {} not implmented
yet.'.format(method))
        # extract batch and labels
        batch, labels = [list(x) for x in zip(*tuple_data)]
        batch_data.extend(batch[:batch_size])
    label_data.extend(labels[:batch_size])
# Trim batch and label at the end
batch_data = batch_data[:batch_size]
label_data = label_data[:batch_size]
# Convert to numpy array
batch_data = np.array(batch_data)
label_data = np.transpose(np.array([label_data]))
return batch_data, label_data
```

11. 初始化嵌入矩阵，声明占位符和嵌入查找函数，代码如下：

```
embeddings = tf.Variable(tf.random_uniform([vocabulary_size,
    embedding_size], -1.0, 1.0))
# Create data/target placeholders
x_inputs = tf.placeholder(tf.int32, shape=[batch_size])
y_target = tf.placeholder(tf.int32, shape=[batch_size, 1])
valid_dataset = tf.constant(valid_examples, dtype=tf.int32)

# Lookup the word embedding:
embed = tf.nn.embedding_lookup(embeddings, x_inputs)
```

12. softmax 损失函数是用来实现多类分类问题常见的损失函数，上一节中其计算预测错误单词分类的损失。因为本例中目标是 10 000 个不同的分类，所以会导致稀疏性非常高。

稀疏性会导致算法模型拟合或者收敛的问题。为了解决该问题，我们使用噪声对比损失函数（noise-contrastive estimation，NCE）。NCE 损失函数将问题转换成一个二值预测，预测单词分类和随机噪声。num_sampled 参数控制转换为随机噪声的批量大小。具体代码如下：

```
nce_weights = tf.Variable(tf.truncated_normal([vocabulary_size,
    embedding_size], stddev=1.0 / np.sqrt(embedding_size)))
nce_biases = tf.Variable(tf.zeros([vocabulary_size]))
loss = tf.reduce_mean(tf.nn.nce_loss(weights=nce_weights,
                                     biases=nce_biases,
                                     inputs=embed,
                                     labels=y_target,
                                     num_sampled=num_sampled,
                                     num_classes=vocabulary_size))
```

13. 创建函数寻找验证单词周围的单词。我们将计算验证单词集和所有词向量之间的余弦相似度，打印出每个验证单词最接近的单词，代码如下：

```
norm = tf.sqrt(tf.reduce_sum(tf.square(embeddings), 1,
    keepdims=True))
normalized_embeddings = embeddings / norm
valid_embeddings = tf.nn.embedding_lookup(normalized_embeddings,
    valid_dataset)
similarity = tf.matmul(valid_embeddings, normalized_embeddings,
    transpose_b=True)
```

14. 声明优化器函数，初始化模型变量，代码如下：

```
optimizer =
tf.train.GradientDescentOptimizer(learning_rate=1.0).minimize(loss)
init = tf.global_variables_initializer()
sess.run(init)
```

15. 训练词嵌入，打印出损失函数和验证单词集单词的最接近的单词，代码如下：

```
loss_vec = []
loss_x_vec = []
for i in range(generations):
    batch_inputs, batch_labels = generate_batch_data(text_data,
batch_size, window_size)
    feed_dict = {x_inputs : batch_inputs, y_target : batch_labels}
    # Run the train step
    sess.run(optimizer, feed_dict=feed_dict)
    # Return the loss
    if (i+1) % print_loss_every == 0:
        loss_val = sess.run(loss, feed_dict=feed_dict)
        loss_vec.append(loss_val)
        loss_x_vec.append(i+1)
        print("Loss at step {} : {}".format(i+1, loss_val))
    # Validation: Print some random words and top 5 related words
    if (i+1) % print_valid_every == 0:
        sim = sess.run(similarity, feed_dict=feed_dict)
        for j in range(len(valid_words)):
            valid_word = word_dictionary_rev[valid_examples[j]]
            top_k = 5 # number of nearest neighbors
            nearest = (-sim[j, :]).argsort()[1:top_k+1]
            log_str = "Nearest to {}:".format(valid_word)
            for k in range(top_k):
```

```
            close_word = word_dictionary_rev[nearest[k]]
            log_str = "%s %s," % (log_str, close_word)
        print(log_str)
```

 以上代码中，使用 argsort 方法进行排序，确保我们找索引时是按照相似度从高到低，而不是相反。

16. 输出结果如下：

```
Loss at step 500 : 13.387781143188477
Loss at step 1000 : 7.240757465362549
Loss at step 49500 : 0.9395825862884521
Loss at step 50000 : 0.30323168635368347
Nearest to cliche: walk, intrigue, brim, eileen, dumber,
Nearest to love: plight, fiction, complete, lady, bartleby,
Nearest to hate: style, throws, players, fearlessness, astringent,
Nearest to silly: delivers, meow, regain, nicely, anger,
Nearest to sad: dizzying, variety, existing, environment, tunney,
```

7.4.3 工作原理

在电影影评文本集上通过 skip-gram 方法训练完 Word2Vec 模型。下载数据集，把单词转换为单词索引，并使用索引数字进行嵌入查找。最后训练预测每个单词最接近的单词。

7.4.4 延伸学习

乍一看，我们期望的验证单词集最接近的单词是同义词。但是这种情况在一个句子中是不常见的，因为一个句子中的每个单词周围的单词都很少出现同义词。我们得到的是数据集中每个单词的同义词的预测。

为了使用单词嵌入，我们必须将其保存和重用。下一节，我们将实现 CBOW 词嵌入模型。

7.5 用 TensorFlow 实现 CBOW 词嵌入模型

本节将实现 word2Vec 的 CBOW 方法。该方法和 skip-gram 方法相似，有一点不同的是，CBOW 根据上下文相关的单词集合来预测目标单词。

7.5.1 开始

在上一个例子中，我们把上下文窗口的单词和目标单词组成输入和输出对，但是在 CBOW 模型中，我们把上下文窗口的单词嵌入放在一起，预测目标单词嵌入，如图 7-5 所示。

大部分实现代码与上一个例子相同，有一点不同的是如何创建单词嵌入和如何从句子中生成嵌入数据。

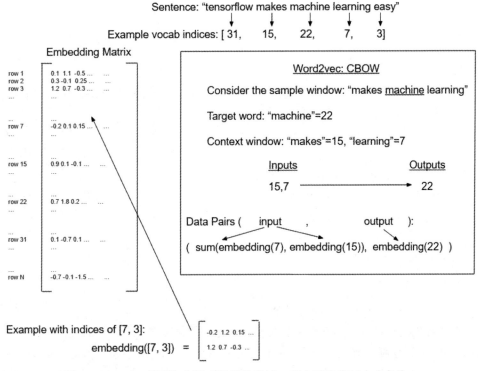

图 7-5 CBOW 模型如何嵌入数据的描述，其中两边窗口大小各为 1

为了让代码更易理解，我们把相关的主要函数移到一个单独的 text_helpers.py 文件中。该文件包含处理数据加载函数、文本归一化函数、单词字典创建函数和批量数据生成函数，这些函数在前面也曾定义。

7.5.2 动手做

1. 导入必要的编程库，包括前面的 text_helpers.py 脚本，该脚本可以进行文本加载和处理。然后创建一个计算图会话，代码如下：

```
import tensorflow as tf
import matplotlib.pyplot as plt
import numpy as np
import random
import os
import pickle
import string
import requests
import collections
import io
import tarfile
import urllib.request
import text_helpers
from nltk.corpus import stopwords
sess = tf.Session()
```

2. 确保临时数据和参数存储在文件夹中，代码如下：

```
# Make a saving directory if it doesn't exist
data_folder_name = 'temp'
if not os.path.exists(data_folder_name):
    os.makedirs(data_folder_name)
```

3. 声明算法模型的参数，这些参数和上一节中的 skip-gram 模型类似，代码如下：

```
# Declare model parameters
batch_size = 500
embedding_size = 200
vocabulary_size = 2000
generations = 50000
model_learning_rate = 0.001
num_sampled = int(batch_size/2
window_size = 3
# Add checkpoints to training
save_embeddings_every = 5000
print_valid_every = 5000
print_loss_every = 100
# Declare stop words
stops = stopwords.words('english')
# We pick some test words. We are expecting synonyms to appear
valid_words = ['love', 'hate', 'happy', 'sad', 'man', 'woman']
```

4. 最初我们已经将数据加载函数和文本归一化函数封装在一个独立的文件中，该文件可由以下Github库得到：https://github.com/ nfmcclure/ tensorflow_ cookbook/ tree/master/ 07_ Natural_Language_ Processing/05_ Working_ With_ CBOW_ Embeddings 和 https://github.com/ PacktPublishing/ TensorFlow-Machine- Learning- Cookbook- Second- Edition，现在可以调用这些函数了。在本例中，设置电影影评不小于三个单词，代码如下：

```
texts, target = text_helpers.load_movie_data(data_folder_name)
texts = text_helpers.normalize_text(texts, stops) # Texts must
contain at least 3 words target = [target[ix] for ix, x in
enumerate(texts) if len(x.split()) > 2] texts = [x for x in texts
if len(x.split()) > 2]
```

5. 创建单词字典，以便查找单词。同时，我们也需要一个逆序单词字典，可以通过索引查找单词。当我们想打印出离验证单词集中单词最近的单词时，可使用逆序单词字典，代码如下：

```
word_dictionary = text_helpers.build_dictionary(texts,
vocabulary_size)
word_dictionary_rev = dict(zip(word_dictionary.values(),
word_dictionary.keys()))
text_data = text_helpers.text_to_numbers(texts, word_dictionary)
# Get validation word keys
valid_examples = [word_dictionary[x] for x in valid_words]
```

6. 初始化待拟合的词嵌入并声明算法模型的数据占位符，代码如下：

```
embeddings = tf.Variable(tf.random_uniform([vocabulary_size,
embedding_size], -1.0, 1.0))
# Create data/target placeholders
x_inputs = tf.placeholder(tf.int32, shape=[batch_size,
```

```
2*window_size])
y_target = tf.placeholder(tf.int32, shape=[batch_size, 1])
valid_dataset = tf.constant(valid_examples, dtype=tf.int32)
```

7. 处理词嵌入。因为 CBOW 模型将上下文窗口内的词嵌入叠加在一起,所以创建一个循环将窗口内的所有词嵌入加在一起,代码如下:

```
# Lookup the word embeddings and
# Add together window embeddings:
embed = tf.zeros([batch_size, embedding_size])
for element in range(2*window_size):
    embed += tf.nn.embedding_lookup(embeddings, x_inputs[:, element])
```

8. 使用 TensorFlow 内建的 NCE 损失函数。这是因为本例输出结果对于 softmax 函数来说太稀疏,导致存在收敛问题,代码如下:

```
# NCE loss parameters
nce_weights = tf.Variable(tf.truncated_normal([vocabulary_size,
embedding_size], stddev=1.0 / np.sqrt(embedding_size)))
nce_biases = tf.Variable(tf.zeros([vocabulary_size]))
# Declare loss function (NCE)
loss = tf.reduce_mean(tf.nn.nce_loss(weights=nce_weights,
                                     biases=nce_biases,
                                     inputs=embed,
                                     labels=y_target,
                                     num_sampled=num_sampled,
                                     num_classes=vocabulary_size))
```

9. 如上一节中的 skip-gram 模型一样,我们使用余弦相似度给出距离验证集单词最近的单词以验证词嵌入质量,代码如下:

```
norm = tf.sqrt(tf.reduce_sum(tf.square(embeddings), 1,
keepdims=True))
normalized_embeddings = embeddings / norm
valid_embeddings = tf.nn.embedding_lookup(normalized_embeddings,
valid_dataset)
similarity = tf.matmul(valid_embeddings, normalized_embeddings,
transpose_b=True)
```

10. 为了保存词向量,我们需要加载 TensorFlow 的 train.Saver() 方法。该方法默认会保存整个计算图会话,但是本例中我们会指定参数只保存嵌入变量,并设置名字。这里设置保存的名字与计算图中的变量名相同,代码如下:

```
saver = tf.train.Saver({"embeddings": embeddings})
```

11. 现在声明优化器函数,初始化模型变量,代码如下:

```
optimizer =
tf.train.GradientDescentOptimizer(learning_rate=model_learning_rate
).minimize(loss)
init = tf.global_variables_initializer()
sess.run(init)
```

12. 最后,遍历迭代训练,打印损失函数,保存词嵌入和字典到指定文件夹,代码如下:

```
loss_vec = []
loss_x_vec = []
for i in range(generations):
    batch_inputs, batch_labels = text_helpers.generate_batch_data(text_data, batch_size, window_size, method='cbow')
    feed_dict = {x_inputs : batch_inputs, y_target : batch_labels}
    # Run the train step
    sess.run(optimizer, feed_dict=feed_dict)
    # Return the loss
    if (i+1) % print_loss_every == 0:
        loss_val = sess.run(loss, feed_dict=feed_dict)
        loss_vec.append(loss_val)
        loss_x_vec.append(i+1)
        print('Loss at step {} : {}'.format(i+1, loss_val))
    # Validation: Print some random words and top 5 related words
    if (i+1) % print_valid_every == 0:
        sim = sess.run(similarity, feed_dict=feed_dict)
        for j in range(len(valid_words)):
            valid_word = word_dictionary_rev[valid_examples[j]]
            top_k = 5 # number of nearest neighbors
            nearest = (-sim[j, :]).argsort()[1:top_k+1]
            log_str = "Nearest to {}:".format(valid_word)
            for k in range(top_k):
                close_word = word_dictionary_rev[nearest[k]]
                print_str = '{} {},'.format(log_str, close_word)
            print(print_str)
    # Save dictionary + embeddings
    if (i+1) % save_embeddings_every == 0:
        # Save vocabulary dictionary
        with open(os.path.join(data_folder_name,'movie_vocab.pkl'),'wb') as f:
            pickle.dump(word_dictionary, f)
        # Save embeddings
        model_checkpoint_path = os.path.join(os.getcwd(),data_folder_name,'cbow_movie_embeddings.ckpt')
        save_path = saver.save(sess, model_checkpoint_path)
        print('Model saved in file: {}'.format(save_path))
```

13. 输出结果如下：

```
Loss at step 100 : 62.04829025268555
Loss at step 200 : 33.182334899902344
...
Loss at step 49900 : 1.6794960498809814
Loss at step 50000 : 1.5071022510528564
Nearest to love: clarity, cult, cliched, literary, memory,
Nearest to hate: bringing, gifted, almost, next, wish,
Nearest to happy: ensemble, fall, courage, uneven, girls,
Nearest to sad: santa, devoid, biopic, genuinely, becomes,
Nearest to man: project, stands, none, soul, away,
Nearest to woman: crush, even, x, team, ensemble,
Model saved in file: .../temp/cbow_movie_embeddings.ckpt
```

14. text_helpers.py 脚本中最后一个函数是上一节的 generate_batch_data() 函数，这里稍微修改一下，增加一个 cbow 方法，代码如下：

```
elif method=='cbow':
    batch_and_labels = [(x[:y] + x[(y+1):], x[y]) for x,y in
zip(window_sequences, label_indices)]
    # Only keep windows with consistent 2*window_size
    batch_and_labels = [(x,y) for x,y in batch_and_labels if
len(x)==2*window_size]
    batch, labels = [list(x) for x in zip(*batch_and_labels)]
```

7.5.3 工作原理

在本节中，Word2Vec 嵌套的 CBOW 模型和 skip-gram 模型非常相似。主要的不同点是生成的数据和处理词嵌入的方式。

本节我们加载文本数据，归一化文本，创建词汇字典，使用词汇字典查找词嵌入，叠加嵌入并训练神经网络模型预测目标单词。

7.5.4 延伸学习

值得注意的是，CBOW 方法是在上下文窗口内叠加词嵌入上进行训练并预测目标单词的。相比于 skip-gram，Word2Vec 的 CBOW 方法更平滑，更适用于小文本数据集。

7.6 使用 TensorFlow 的 Word2Vec 预测

本节将使用前面学过的词嵌入策略进行分类。

7.6.1 开始

现在已经创建并保存了 CBOW 词嵌入，我们使用它们在电影影评数据集上做情感分析。本节将介绍如何加载和使用预训练的嵌入，并使用这些词嵌入进行情感分析，通过训练线性逻辑回归模型来预测电影的好坏。

情感分析是一个相当棘手的任务，因为很难捕捉到人类语言真实含义的微妙变化和细微区别。挖苦、玩笑和模棱两可的含义使得任务异常艰难。我们将创建一个线性逻辑回归模型训练电影影评数据集，来看是否可以得到上一节 CBOW 模型所创建并保存的词嵌入之外的信息。因为本节关注的是加载和使用预训练好的词嵌入，所以这里尽量不使用复杂算法模型。

7.6.2 动手做

1. 导入必要的编程库和计算图会话，代码如下：

```
import tensorflow as tf
import matplotlib.pyplot as plt
import numpy as np
import random
import os
import pickle
import string
```

```
import requests
import collections
import io
import tarfile
import urllib.request
import text_helpers
from nltk.corpus import stopwords
sess = tf.Session()
```

2. 声明算法模型参数。应选择与前一节中 CBOW 方法相同的嵌入大小，代码如下：

```
embedding_size = 200
vocabulary_size = 2000
batch_size = 100
max_words = 100
stops = stopwords.words('english')
```

3. 用 text_helpers.py 脚本加载和转换文本数据集，代码如下：

```
texts, target = text_helpers.load_movie_data()
# Normalize text
print('Normalizing Text Data')
texts = text_helpers.normalize_text(texts, stops)
# Texts must contain at least 3 words
target = [target[ix] for ix, x in enumerate(texts) if
len(x.split()) > 2]
texts = [x for x in texts if len(x.split()) > 2]
train_indices = np.random.choice(len(target),
round(0.8*len(target)), replace=False)
test_indices = np.array(list(set(range(len(target))) -
set(train_indices)))
texts_train = [x for ix, x in enumerate(texts) if ix in
train_indices]
texts_test = [x for ix, x in enumerate(texts) if ix in
test_indices]
target_train = np.array([x for ix, x in enumerate(target) if ix in
train_indices])
target_test = np.array([x for ix, x in enumerate(target) if ix in
test_indices])
```

4. 现在加载 CBOW 嵌入中保存的单词字典，这一步很重要，使得我们拥有相同的单词到嵌套索引的映射，代码如下：

```
dict_file = os.path.join(data_folder_name, 'movie_vocab.pkl')
word_dictionary = pickle.load(open(dict_file, 'rb'))
```

5. 通过单词字典将加载的句子转化为数值型 numpy 数组，代码如下：

```
text_data_train =
np.array(text_helpers.text_to_numbers(texts_train,
word_dictionary))
text_data_test = np.array(text_helpers.text_to_numbers(texts_test,
word_dictionary))
```

6. 由于电影影评长度不一，我们用同一长度（设为 100 个单词长度）将其标准化。如果电影影评长度少于 100 个单词，我们将用 0 去填充，代码如下：

```
text_data_train = np.array([x[0:max_words] for x in
[y+[0]*max_words for y in text_data_train]])
```

```
text_data_test = np.array([x[0:max_words] for x in [y+[0]*max_words
for y in text_data_test]])
```

7. 声明逻辑回归的模型变量和占位符，代码如下：

```
A = tf.Variable(tf.random_normal(shape=[embedding_size,1]))
b = tf.Variable(tf.random_normal(shape=[1,1]))
# Initialize placeholders
x_data = tf.placeholder(shape=[None, max_words], dtype=tf.int32)
y_target = tf.placeholder(shape=[None, 1], dtype=tf.float32)
```

8. 为了使得TensorFlow可以重用训练过的词向量，首先需要给存储方法设置一个变量。这里创建一个嵌入变量，其形状与将要加载的嵌入相同，代码如下：

```
embeddings = tf.Variable(tf.random_uniform([vocabulary_size,
embedding_size], -1.0, 1.0))
```

9. 在计算图中加入embedding_lookup操作，计算句子中所有单词的平均嵌入，代码如下：

```
embed = tf.nn.embedding_lookup(embeddings, x_data)
# Take average of all word embeddings in documents
embed_avg = tf.reduce_mean(embed, 1)
```

10. 声明模型操作和损失函数。记住，损失函数中已经内建了sigmoid操作，代码如下：

```
model_output = tf.add(tf.matmul(embed_avg, A), b)
# Declare loss function (Cross Entropy loss)
loss =
tf.reduce_mean(tf.nn.sigmoid_cross_entropy_with_logits(logits=model
_output, labels=y_target))
```

11. 在计算图中增加预测函数和准确度函数，评估训练模型的准确度，代码如下：

```
prediction = tf.round(tf.sigmoid(model_output))
predictions_correct = tf.cast(tf.equal(prediction, y_target),
tf.float32)
accuracy = tf.reduce_mean(predictions_correct)
```

12. 声明优化器函数，并初始化下面的模型变量，代码如下：

```
my_opt = tf.train.AdagradOptimizer(0.005)
train_step = my_opt.minimize(loss)
init = tf.global_variables_initializer()
sess.run(init)
```

13. 随机初始化嵌入后，调用Saver方法来加载上一节中保存好的CBOW嵌入到嵌入变量，代码如下：

```
model_checkpoint_path =
os.path.join(data_folder_name,'cbow_movie_embeddings.ckpt')
saver = tf.train.Saver({"embeddings": embeddings})
saver.restore(sess, model_checkpoint_path)
```

14. 开始迭代训练。注意，每迭代100次就保存训练集和测试集的损失和准确度。每迭代500次就打印一次模型状态，代码如下：

```
train_loss = []
test_loss = []
train_acc = []
test_acc = []
```

```
i_data = []
for i in range(10000):
    rand_index = np.random.choice(text_data_train.shape[0],
size=batch_size)
    rand_x = text_data_train[rand_index]
    rand_y = np.transpose([target_train[rand_index]])
    sess.run(train_step, feed_dict={x_data: rand_x, y_target:
rand_y})
    # Only record loss and accuracy every 100 generations
    if (i+1)%100==0:
        i_data.append(i+1)
        train_loss_temp = sess.run(loss, feed_dict={x_data: rand_x,
y_target: rand_y})
        train_loss.append(train_loss_temp)
        test_loss_temp = sess.run(loss, feed_dict={x_data:
text_data_test, y_target: np.transpose([target_test])})
        test_loss.append(test_loss_temp)
        train_acc_temp = sess.run(accuracy, feed_dict={x_data:
rand_x, y_target: rand_y})
        train_acc.append(train_acc_temp)
        test_acc_temp = sess.run(accuracy, feed_dict={x_data:
text_data_test, y_target: np.transpose([target_test])})
        test_acc.append(test_acc_temp)
    if (i+1)%500==0:
        acc_and_loss = [i+1, train_loss_temp, test_loss_temp,
train_acc_temp, test_acc_temp]
        acc_and_loss = [np.round(x,2) for x in acc_and_loss]
        print('Generation # {}. Train Loss (Test Loss): {:.2f}
({:.2f}). Train Acc (Test Acc): {:.2f}
({:.2f})'.format(*acc_and_loss))
```

15. 打印结果如下：

Generation # 500. Train Loss (Test Loss): 0.70 (0.71). Train Acc (Test Acc): 0.52 (0.48)
Generation # 1000. Train Loss (Test Loss): 0.69 (0.72). Train Acc (Test Acc): 0.56 (0.47)
...
Generation # 9500. Train Loss (Test Loss): 0.69 (0.70). Train Acc (Test Acc): 0.57 (0.55)
Generation # 10000. Train Loss (Test Loss): 0.70 (0.70). Train Acc (Test Acc): 0.59 (0.55)

16. 每迭代100次就绘制训练集和测试集损失函数和准确度，代码如下：

```
# Plot loss over time
plt.plot(i_data, train_loss, 'k-', label='Train Loss')
plt.plot(i_data, test_loss, 'r--', label='Test Loss', linewidth=4)
# Plot loss over time
plt.plot(i_data, train_loss, 'k-', label='Train Loss')
plt.plot(i_data, test_loss, 'r--', label='Test Loss', linewidth=4)
plt.title('Cross Entropy Loss per Generation')
plt.xlabel('Generation')
plt.ylabel('Cross Entropy Loss')
plt.legend(loc='upper right')
plt.show()

# Plot train and test accuracy
```

```
plt.plot(i_data, train_acc, 'k-', label='Train Set Accuracy')
plt.plot(i_data, test_acc, 'r--', label='Test Set Accuracy',
linewidth=4)
plt.title('Train and Test Accuracy')
plt.xlabel('Generation')
plt.ylabel('Accuracy')
plt.legend(loc='lower right')
plt.show()
```

交叉熵损失如图 7-6 所示，训练和测试准确度如图 7-7 所示。

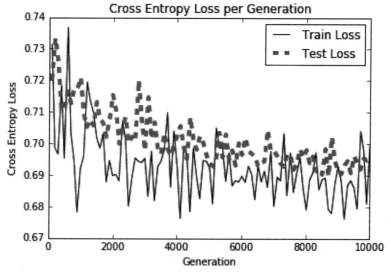

图 7-6　迭代 10 000 次的训练集和测试集损失图

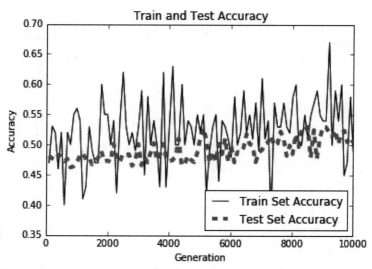

图 7-7　迭代 10 000 次的训练集准确度和测试集准确度趋势图，两者准确度均在缓慢提高。值得注意的是，该模型性能较差，仅仅比随机预测好那么一点点

7.6.3 工作原理

我们加载前一节的CBOW嵌入,并在电影影评的平均嵌入上进行逻辑回归模型训练。本节的关注点是,如何从磁盘上加载模型变量到已经初始化的当前模型变量中,以及如何存储和加载已创建的单词字典来训练嵌入。当使用相同的嵌入时,需要使用相同的单词到嵌入索引的映射。

7.6.4 延伸学习

我们获得了60%的情感预测准确度。例如,预知单词great背后的含义是一件非常困难的任务,该单词可以用在正面或者负面的电影影评中。

为了解决该问题,我们计划针对文档本身创建嵌入来解决情感问题。通常,一个完整的影评要么是正面的,要么是负面的,我们可以以此来解决情感分析问题。

7.7 用TensorFlow实现基于Doc2Vec的情感分析

现在知道如何训练词嵌入了,我们可以扩展这些方法使其成为文档嵌入。本节将使用TensorFlow实现基于Doc2Vec的情感分析。

7.7.1 开始

在上一节关于Word2Vec的方法中,我们处理的是单词之间的位置关系。但是没有考虑单词和单词所在文档或影评之间的关系。Word2Vec方法的扩展之一就是Doc2Vec方法,其考虑文档的影响。

Doc2Vec方法的基本思想是引入文档嵌入,同时连同单词嵌入一起帮助判断文档的感情色彩。例如,如果单词movie和love接连出现,那也不能帮助我们判断该电影影评的情感。影评可能谈论他们如何喜爱电影,也可能谈论他们如何讨厌电影。但是,如果电影影评有足够的长度,并且包含多个负面词汇,那么文档整体的感情色彩可以帮助我们预测最近的单词。

Doc2Vec方法只是简单地增加了一个文档嵌入矩阵,并使用单词的窗口叠加文档索引进行单词预测。在一个文档中的所有单词窗口有相同的文档索引。值得一提的是,如何结合文档嵌入和单词嵌入是讨论重点。我们将单词窗口内的单词嵌入求和,主要有两种结合这些单词嵌入和文档嵌入的方法。一般将文档嵌入和单词嵌入相加,或者将文档嵌套直接追加在单词嵌入后面。如果将两种嵌入相加,那么要求文档嵌入的大小和单词嵌入大小相同;如果将两种嵌入直接追加,那么需要增加逻辑回归模型的变量数量。本节我们将讲解如何处理连接两种嵌入。不过一般来讲,对于小数据集,两种嵌入相加的方法是更好的选择。

首先进行电影影评的文档嵌入和单词嵌入,然后分割为训练集和测试集,训练逻辑回归模型,并评估该方法是否可以提高预测影评情感的准确度。

7.7.2 动手做

1. 导入必要的编程库，开始一个计算图会话，代码如下：

```
import tensorflow as tf
import matplotlib.pyplot as plt
import numpy as np
import random
import os
import pickle
import string
import requests
import collections
import io
import tarfile
import urllib.request
import text_helpers
from nltk.corpus import stopwords
sess = tf.Session()
```

2. 加载影评数据集，代码如下：

```
texts, target = text_helpers.load_movie_data()
```

3. 声明算法模型参数，代码如下：

```
batch_size = 500
vocabulary_size = 7500
generations = 100000
model_learning_rate = 0.001
embedding_size = 200    # Word embedding size
doc_embedding_size = 100   # Document embedding size
concatenated_size = embedding_size + doc_embedding_size
num_sampled = int(batch_size/2)
window_size = 3         # How many words to consider to the left.
# Add checkpoints to training
save_embeddings_every = 5000
print_valid_every = 5000
print_loss_every = 100
# Declare stop words
stops = stopwords.words('english')
# We pick a few test words.
valid_words = ['love', 'hate', 'happy', 'sad', 'man', 'woman']
```

4. 归一化电影影评，确保每条影评都大于指定的窗口大小，代码如下：

```
texts = text_helpers.normalize_text(texts, stops)
# Texts must contain at least as much as the prior window size
target = [target[ix] for ix, x in enumerate(texts) if len(x.split()) > window_size]
texts = [x for x in texts if len(x.split()) > window_size]
assert(len(target)==len(texts))
```

5. 创建单词字典。值得注意的是，我们无须创建文档字典。文档索引仅仅是文档的索引值，每个文档有唯一的索引值，代码如下：

```
word_dictionary = text_helpers.build_dictionary(texts, vocabulary_size)
```

```
word_dictionary_rev = dict(zip(word_dictionary.values(),
word_dictionary.keys()))
text_data = text_helpers.text_to_numbers(texts, word_dictionary)
# Get validation word keys
valid_examples = [word_dictionary[x] for x in valid_words]
```

6. 定义单词嵌套和文档嵌套。然后声明 NCE 损失函数参数，代码如下：

```
embeddings = tf.Variable(tf.random_uniform([vocabulary_size,
embedding_size], -1.0, 1.0))
doc_embeddings = tf.Variable(tf.random_uniform([len(texts),
doc_embedding_size], -1.0, 1.0))
# NCE loss parameters
nce_weights = tf.Variable(tf.truncated_normal([vocabulary_size,
concatenated_size],
                                              stddev=1.0 /
np.sqrt(concatenated_size)))
nce_biases = tf.Variable(tf.zeros([vocabulary_size]))
```

7. 声明 Doc2Vec 索引和目标单词索引的占位符。注意，输入索引的大小是窗口大小加1，这是因为每个生成的数据窗口将有一个额外的文档索引，代码如下：

```
x_inputs = tf.placeholder(tf.int32, shape=[None, window_size + 1])
y_target = tf.placeholder(tf.int32, shape=[None, 1])
valid_dataset = tf.constant(valid_examples, dtype=tf.int32)
```

8. 创建嵌入函数将单词嵌入求和，然后连接文档嵌入，代码如下：

```
embed = tf.zeros([batch_size, embedding_size])
for element in range(window_size):
    embed += tf.nn.embedding_lookup(embeddings, x_inputs[:,
element])
doc_indices = tf.slice(x_inputs, [0,window_size],[batch_size,1])
doc_embed = tf.nn.embedding_lookup(doc_embeddings,doc_indices)
# concatenate embeddings
final_embed = tf.concat(axis=1, values=)
```

9. 现在声明损失函数并创建优化器，代码如下：

```
loss = tf.reduce_mean(tf.nn.nce_loss(weights=nce_weights,
                                     biases=nce_biases,
                                     labels=y_target,
                                     inputs=final_embed,
                                     num_sampled=num_sampled,
                                     num_classes=vocabulary_size))
# Create optimizer
optimizer =
tf.train.GradientDescentOptimizer(learning_rate=model_learning_rate
)
train_step = optimizer.minimize(loss)
```

10. 声明验证单词集的余弦距离，代码如下：

```
norm = tf.sqrt(tf.reduce_sum(tf.square(embeddings), 1,
keep_dims=True))
normalized_embeddings = embeddings / norm
valid_embeddings = tf.nn.embedding_lookup(normalized_embeddings,
valid_dataset)
similarity = tf.matmul(valid_embeddings, normalized_embeddings,
transpose_b=True)
```

11. 为了保存嵌入，创建模型的 saver 函数，然后初始化模型变量，代码如下：

```
saver = tf.train.Saver({"embeddings": embeddings, "doc_embeddings":
doc_embeddings})
init = tf.global_variables_initializer()
sess.run(init)
loss_vec = []
loss_x_vec = []
for i in range(generations):
    batch_inputs, batch_labels =
text_helpers.generate_batch_data(text_data, batch_size,
window_size, method='doc2vec')
    feed_dict = {x_inputs : batch_inputs, y_target : batch_labels}

    # Run the train step
    sess.run(train_step, feed_dict=feed_dict)

    # Return the loss
    if (i+1) % print_loss_every == 0:
        loss_val = sess.run(loss, feed_dict=feed_dict)
        loss_vec.append(loss_val)
        loss_x_vec.append(i+1)
        print('Loss at step {} : {}'.format(i+1, loss_val))
    # Validation: Print some random words and top 5 related words
    if (i+1) % print_valid_every == 0:
        sim = sess.run(similarity, feed_dict=feed_dict)
        for j in range(len(valid_words)):
            valid_word = word_dictionary_rev[valid_examples[j]]
            top_k = 5 # number of nearest neighbors
            nearest = (-sim[j, :]).argsort()[1:top_k+1]
            log_str = "Nearest to {}:".format(valid_word)
            for k in range(top_k):
                close_word = word_dictionary_rev[nearest[k]]
                log_str = '{} {},'.format(log_str, close_word)
            print(log_str)
    # Save dictionary + embeddings
    if (i+1) % save_embeddings_every == 0:
        # Save vocabulary dictionary
        with open(os.path.join(data_folder_name,'movie_vocab.pkl'),
'wb') as f:
            pickle.dump(word_dictionary, f)
        # Save embeddings
        model_checkpoint_path =
os.path.join(os.getcwd(),data_folder_name,'doc2vec_movie_embeddings
.ckpt')
        save_path = saver.save(sess, model_checkpoint_path)
        print('Model saved in file: {}'.format(save_path))
```

12. 打印结果如下：

```
Loss at step 100 : 126.176816940307617
Loss at step 200 : 89.608322143554688
...
Loss at step 99900 : 17.733346939086914
Loss at step 100000 : 17.384489059448242
Nearest to love: ride, with, by, its, start,
Nearest to hate: redundant, snapshot, from, performances,
extravagant,
```

```
Nearest to happy: queen, chaos, them, succumb, elegance,
Nearest to sad: terms, pity, chord, wallet, morality,
Nearest to man: of, teen, an, our, physical,
Nearest to woman: innocuous, scenes, prove, except, lady,
Model saved in file: /.../temp/doc2vec_movie_embeddings.ckpt
```

13. 训练完Doc2Vec嵌入，我们能使用这些嵌入训练逻辑回归模型，预测影评情感色彩。首先设置逻辑回归模型的一些参数，代码如下：

```
max_words = 20 # maximum review word length
logistic_batch_size = 500 # training batch size
```

14. 分割数据集为训练集和测试集，代码如下：

```
train_indices = np.sort(np.random.choice(len(target),
round(0.8*len(target)), replace=False))
test_indices = np.sort(np.array(list(set(range(len(target))) -
set(train_indices))))
texts_train = [x for ix, x in enumerate(texts) if ix in
train_indices]
texts_test = [x for ix, x in enumerate(texts) if ix in
test_indices]
target_train = np.array([x for ix, x in enumerate(target) if ix in
train_indices])
target_test = np.array([x for ix, x in enumerate(target) if ix in
test_indices])
```

15. 将电影影评转换成数值型的单词索引，填充或者裁剪每条影评为20个单词，代码如下：

```
text_data_train =
np.array(text_helpers.text_to_numbers(texts_train,
word_dictionary)) text_data_test =
np.array(text_helpers.text_to_numbers(texts_test, word_dictionary))
# Pad/crop movie reviews to specific length text_data_train =
np.array([x[0:max_words] for x in [y+[0]*max_words for y in
text_data_train]]) text_data_test = np.array([x[0:max_words] for x
in [y+[0]*max_words for y in text_data_test]])
```

16. 声明逻辑回归模型的数据占位符、模型变量、模型操作和损失函数，代码如下：

```
# Define Logistic placeholders
log_x_inputs = tf.placeholder(tf.int32, shape=[None, max_words +
1])
log_y_target = tf.placeholder(tf.int32, shape=[None, 1])
A = tf.Variable(tf.random_normal(shape=[concatenated_size,1]))
b = tf.Variable(tf.random_normal(shape=[1,1]))

# Declare logistic model (sigmoid in loss function)
model_output = tf.add(tf.matmul(log_final_embed, A), b)

# Declare loss function (Cross Entropy loss)
logistic_loss =
tf.reduce_mean(tf.nn.sigmoid_cross_entropy_with_logits(logits=model
_output,
labels=tf.cast(log_y_target, tf.float32)))
```

17. 创建另外一个嵌入函数。前半部分的嵌入函数是训练3个单词的窗口和文档索引以预测下一个单词。这里也是类似的功能，不同的是训练20个单词的影评，代码如下：

```
# Add together element embeddings in window:
log_embed = tf.zeros([logistic_batch_size, embedding_size])
for element in range(max_words):
    log_embed += tf.nn.embedding_lookup(embeddings, log_x_inputs[:, element])
log_doc_indices = tf.slice(log_x_inputs, [0,max_words],[logistic_batch_size,1])
log_doc_embed = tf.nn.embedding_lookup(doc_embeddings,log_doc_indices)
# concatenate embeddings
log_final_embed = tf.concat(1, [log_embed, tf.squeeze(log_doc_embed)])
```

18. 创建预测函数和准确度函数，评估迭代训练模型。然后声明优化器函数，初始化所有模型变量，代码如下：

```
prediction = tf.round(tf.sigmoid(model_output))
predictions_correct = tf.cast(tf.equal(prediction, tf.cast(log_y_target, tf.float32)), tf.float32)
accuracy = tf.reduce_mean(predictions_correct)
# Declare optimizer
logistic_opt = tf.train.GradientDescentOptimizer(learning_rate=0.01)
logistic_train_step = logistic_opt.minimize(logistic_loss, var_list=[A, b])
# Intitialize Variables
init = tf.global_variables_initializer()
sess.run(init)
```

19. 开始训练逻辑回归模型，代码如下：

```
train_loss = []
test_loss = []
train_acc = []
test_acc = []
i_data = []
for i in range(10000):
    rand_index = np.random.choice(text_data_train.shape[0], size=logistic_batch_size)
    rand_x = text_data_train[rand_index]
    # Append review index at the end of text data
    rand_x_doc_indices = train_indices[rand_index]
    rand_x = np.hstack((rand_x, np.transpose([rand_x_doc_indices])))
    rand_y = np.transpose([target_train[rand_index]])
    feed_dict = {log_x_inputs : rand_x, log_y_target : rand_y}
    sess.run(logistic_train_step, feed_dict=feed_dict)
    # Only record loss and accuracy every 100 generations
    if (i+1)%100==0:
        rand_index_test = np.random.choice(text_data_test.shape[0], size=logistic_batch_size)
        rand_x_test = text_data_test[rand_index_test]
        # Append review index at the end of text data
        rand_x_doc_indices_test = test_indices[rand_index_test]
        rand_x_test = np.hstack((rand_x_test, np.transpose([rand_x_doc_indices_test])))
```

```
        rand_y_test = np.transpose([target_test[rand_index_test]])
        test_feed_dict = {log_x_inputs: rand_x_test, log_y_target:
rand_y_test}
        i_data.append(i+1)
        train_loss_temp = sess.run(logistic_loss,
feed_dict=feed_dict)
        train_loss.append(train_loss_temp)
        test_loss_temp = sess.run(logistic_loss,
feed_dict=test_feed_dict)
        test_loss.append(test_loss_temp)
        train_acc_temp = sess.run(accuracy, feed_dict=feed_dict)
        train_acc.append(train_acc_temp)
        test_acc_temp = sess.run(accuracy,
feed_dict=test_feed_dict)
        test_acc.append(test_acc_temp)
    if (i+1)%500==0:
        acc_and_loss = [i+1, train_loss_temp, test_loss_temp,
train_acc_temp, test_acc_temp]
        acc_and_loss = [np.round(x,2) for x in acc_and_loss]
        print('Generation # {}. Train Loss (Test Loss): {:.2f}
({:.2f}). Train Acc (Test Acc): {:.2f}
({:.2f})'.format(*acc_and_loss))
```

20. 打印结果如下：

Generation # 500. Train Loss (Test Loss): 5.62 (7.45). Train Acc (Test Acc): 0.52 (0.48) Generation # 10000. Train Loss (Test Loss): 2.35 (2.51). Train Acc (Test Acc): 0.59 (0.58)

21. 我们已经创建了独立的批量数据生成的方法——text_helpers.generate_batch_data() 函数。本节前面使用该方法训练 Doc2Vec 嵌入，以下代码取自该函数，属于该方法：

```
def generate_batch_data(sentences, batch_size, window_size,
method='skip_gram'):
    # Fill up data batch
    batch_data = []
    label_data = []
    while len(batch_data) < batch_size:
        # select random sentence to start
        rand_sentence_ix = int(np.random.choice(len(sentences),
size=1))
        rand_sentence = sentences[rand_sentence_ix]
        # Generate consecutive windows to look at
        window_sequences = [rand_sentence[max((ix-
window_size),0):(ix+window_size+1)] for ix, x in
enumerate(rand_sentence)]
        # Denote which element of each window is the center word of
interest
        label_indices = [ix if ix<window_size else window_size for
ix,x in enumerate(window_sequences)]
        # Pull out center word of interest for each window and
create a tuple for each window
        if method=='skip_gram':
            ...
        elif method=='cbow':
            ...
        elif method=='doc2vec':
```

```
            # For doc2vec we keep LHS window only to predict target
word
            batch_and_labels = [(rand_sentence[i:i+window_size],
rand_sentence[i+window_size]) for i in range(0, len(rand_sentence)-
window_size)]
            batch, labels = [list(x) for x in
zip(*batch_and_labels)]
            # Add document index to batch!! Remember that we must
extract the last index in batch for the doc-index
            batch = [x + [rand_sentence_ix] for x in batch]
        else:
            raise ValueError('Method {} not implmented
yet.'.format(method))
        # extract batch and labels
        batch_data.extend(batch[:batch_size])
        label_data.extend(labels[:batch_size])
# Trim batch and label at the end
batch_data = batch_data[:batch_size]
label_data = label_data[:batch_size]
# Convert to numpy array
batch_data = np.array(batch_data)
label_data = np.transpose(np.array([label_data]))
return batch_data, label_data
```

7.7.3 工作原理

在本节中，我们进行了两个迭代训练。第一个是训练 Doc2Vec 嵌入，第二个是在电影影评上训练逻辑回归模型预测影评情感色彩。

虽然情感分析预测的准确度没有增加多少（仍然约为 60%），但是我们成功实现了电影影评数据集的 Doc2Vec 的连接（级联）版本。因为逻辑回归模型不能捕捉到自然语言处理中的非线性行为特征，所以为了提高预测的准确度，我们可以尝试用不同的参数训练 Doc2Vec 嵌套，也可以使用稍微复杂一些的算法模型。

CHAPTER 8

第 8 章

卷积神经网络

在过去的几年中，卷积神经网络（Convolutional Neural Networks，CNN）在图像识别方面有了重大突破。本章主要包含以下知识点：
- TensorFlow 实现简单的 CNN
- TensorFlow 实现进阶的 CNN
- 再训练已有的 CNN 模型
- TensorFlow 实现图像风格迁移
- TensorFlow 实现 DeepDream

值得注意的是，读者可以在 GitHub 上找到所有代码，网址为：https://github.com/nfmcclure/tensorflow_cookbook 或 https://github.com/PacktPublishing/TensorFlow-Machine-Learning-Cookbook-Second-Edition。

8.1 简介

从数学意义上讲，卷积是一个函数在另外一个函数中的操作。在本例中，我们将在图像中应用矩阵乘法过滤器。为了方便，将图像认为是数字矩阵，这些数字代表像素或图像属性。卷积操作将固定窗口大小的过滤器在图像上滑动，并进行对应位置元素相乘之后取和，便得到卷积结果。图 8-1 展示了图像卷积如何工作。

卷积神经网络也有些必要的操作，比如引入非线性（ReLU 函数），或者聚合参数（maxpool 函数），以及其他相似的操作。前面的图像是在一个 5×5 的数组上进行卷积操作的例子，其中卷积过滤器为 2×2 的矩阵，步长为 1，且仅考虑有效位置。可训练的变量是该操作中 2×2 的过滤器权重。一般紧跟着卷积之后的是聚合操作，比如最大池化。图 8-2 提供一个最大池化操作的例子，其在 2×2 的区域上取最大值，两个方向上的移动步长均为 2。

虽然我们准备创建自定义的 CNN 来进行图像识别，但是，强烈推荐读者使用现有的架构方案，我们在本章后续部分也会使用。

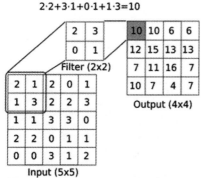
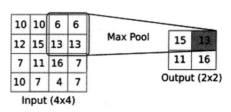
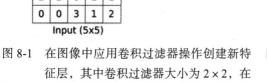

图 8-1 在图像中应用卷积过滤器操作创建新特征层，其中卷积过滤器大小为 2×2，在 5×5 的输入图像的两个方向上以步长为 1 进行卷积。结果是一个 4×4 矩阵

图 8-2 最大池化操作的例子。在本例中，窗口大小为 2×2，在 4×4 的输入（有效位置）的两个方向上以步长为 2 进行操作。结果是 2×2 的矩阵

 一般采用预训练网络，并在最后用全连接在新的数据集上再训练预训练的神经网络。该方法是非常有用的，在 8.4 节也会详细讲述，并再训练已存在的架构来提升 CIFAR-10 的预测效果。

8.2 用 TensorFlow 实现简单的 CNN

本节将开发一个四层卷积神经网络，提升预测 MNIST 数字的准确度。前两个卷积层由 Convolution-ReLU-maxpool 操作组成，后两层是全联接层。

8.2.1 开始

为了访问 MNIST 数据集，TensorFlow 的 examples.tutorials 包包含数据加载功能。数据集加载之后，我们设置算法模型变量，创建模型，批量训练模型，并且可视化损失函数、准确度和一些抽样数字。

8.2.2 动手做

1. 加载必要的编程库，开始计算图会话，代码如下：

```
import matplotlib.pyplot as plt
import numpy as np
import tensorflow as tf
from tensorflow.examples.tutorials.mnist import input_data
from tensorflow.python.framework import ops
ops.reset_default_graph()

sess = tf.Session()
```

2. 加载数据集，转化图像为 28×28 的数组，代码如下：

```
data_dir = 'temp'
mnist = input_data.read_data_sets(data_dir, one_hot=False)
train_xdata = np.array([np.reshape(x, (28,28)) for x in
mnist.train.images])
test_xdata = np.array([np.reshape(x, (28,28)) for x in
mnist.test.images])
train_labels = mnist.train.labels
test_labels = mnist.test.labels
```

注意，下载的 MNIST 数据集包括验证数据集。验证数据集的大小与测试数据集相同。如果我们进行超参数调优或者模型选择，也需要加载测试数据集和验证数据集。

3. 设置模型参数。由于图像是灰度图，所以该图像的深度为 1，即颜色通道数为 1，代码如下：

```
batch_size = 100
learning_rate = 0.005
evaluation_size = 500
image_width = train_xdata[0].shape[0]
image_height = train_xdata[0].shape[1]
target_size = max(train_labels) + 1
num_channels = 1
generations = 500
eval_every = 5
conv1_features = 25
conv2_features = 50
max_pool_size1 = 2
max_pool_size2 = 2
fully_connected_size1 = 100
```

4. 为数据集声明占位符。同时，声明训练数据集变量和测试数据集变量。本例中的训练批量大小和评估大小可以根据实际训练和评估的机器物理内存来调整，代码如下：

```
x_input_shape = (batch_size, image_width, image_height,
num_channels)
x_input = tf.placeholder(tf.float32, shape=x_input_shape)
y_target = tf.placeholder(tf.int32, shape=(batch_size))
eval_input_shape = (evaluation_size, image_width, image_height,
num_channels)
eval_input = tf.placeholder(tf.float32, shape=eval_input_shape)
eval_target = tf.placeholder(tf.int32, shape=(evaluation_size))
```

5. 声明卷积层的权重和偏置，权重和偏置的参数在前面的步骤中已设置过，代码如下：

```
conv1_weight = tf.Variable(tf.truncated_normal([4, 4, num_channels,
conv1_features], stddev=0.1, dtype=tf.float32))
conv1_bias =
tf.Variable(tf.zeros([conv1_features],dtype=tf.float32))
conv2_weight = tf.Variable(tf.truncated_normal([4, 4,
conv1_features, conv2_features], stddev=0.1, dtype=tf.float32))
conv2_bias =
tf.Variable(tf.zeros([conv2_features],dtype=tf.float32))
```

6. 声明全联接层的权重和偏置，代码如下：

```
resulting_width = image_width // (max_pool_size1 * max_pool_size2)
resulting_height = image_height // (max_pool_size1 * max_pool_size2)
full1_input_size = resulting_width * resulting_height*conv2_features
full1_weight = tf.Variable(tf.truncated_normal([full1_input_size, fully_connected_size1], stddev=0.1, dtype=tf.float32))
full1_bias = tf.Variable(tf.truncated_normal([fully_connected_size1], stddev=0.1, dtype=tf.float32))
full2_weight = tf.Variable(tf.truncated_normal([fully_connected_size1, target_size], stddev=0.1, dtype=tf.float32))
full2_bias = tf.Variable(tf.truncated_normal([target_size], stddev=0.1, dtype=tf.float32))
```

7. 声明算法模型。首先，创建一个模型函数 my_conv_net()，注意该函数能访问所有的层权重和偏置。当然，为了最后两层全联接层能有效工作，我们将第二个卷积层的结果摊平，代码如下：

```
def my_conv_net(input_data):
    # First Conv-ReLU-MaxPool Layer
    conv1 = tf.nn.conv2d(input_data, conv1_weight, strides=[1, 1, 1, 1], padding='SAME')
    relu1 = tf.nn.relu(tf.nn.bias_add(conv1, conv1_bias))
    max_pool1 = tf.nn.max_pool(relu1, ksize=[1, max_pool_size1, max_pool_size1, 1], strides=[1, max_pool_size1, max_pool_size1, 1], padding='SAME')
    # Second Conv-ReLU-MaxPool Layer
    conv2 = tf.nn.conv2d(max_pool1, conv2_weight, strides=[1, 1, 1, 1], padding='SAME')
    relu2 = tf.nn.relu(tf.nn.bias_add(conv2, conv2_bias))
    max_pool2 = tf.nn.max_pool(relu2, ksize=[1, max_pool_size2, max_pool_size2, 1], strides=[1, max_pool_size2, max_pool_size2, 1], padding='SAME')
    # Transform Output into a 1xN layer for next fully connected layer
    final_conv_shape = max_pool2.get_shape().as_list()
    final_shape = final_conv_shape[1] * final_conv_shape[2] * final_conv_shape[3]
    flat_output = tf.reshape(max_pool2, [final_conv_shape[0], final_shape])
    # First Fully Connected Layer
    fully_connected1 = tf.nn.relu(tf.add(tf.matmul(flat_output, full1_weight), full1_bias))
    # Second Fully Connected Layer
    final_model_output = tf.add(tf.matmul(fully_connected1, full2_weight), full2_bias)
    return final_model_output
```

8. 声明训练模型和测试模型，代码如下：

```
model_output = my_conv_net(x_input)
test_model_output = my_conv_net(eval_input)
```

9. 因为本例的预测结果不是多分类，而仅仅是一类，所以使用 sparse_softmax 函数作为损失函数代码如下：

```
loss = tf.reduce_mean(tf.nn.sparse_softmax_cross_entropy_with_logits(logits=model_output, labels=y_target))
```

10. 创建训练集和测试集的预测函数。同时，创建对应的准确度函数，评估模型在每批量上的准确度，代码如下：

```
prediction = tf.nn.softmax(model_output)
test_prediction = tf.nn.softmax(test_model_output)
# Create accuracy function
def get_accuracy(logits, targets):
    batch_predictions = np.argmax(logits, axis=1)
    num_correct = np.sum(np.equal(batch_predictions, targets))
    return 100. * num_correct/batch_predictions.shape[0]
```

11. 创建优化器函数，声明训练步长，初始化所有的模型变量，代码如下：

```
my_optimizer = tf.train.MomentumOptimizer(learning_rate, 0.9)
train_step = my_optimizer.minimize(loss)
# Initialize Variables
init = tf.global_variables_initializer()
sess.run(init)
```

12. 开始训练模型。遍历迭代随机选择批量数据进行训练。我们在训练集批量数据和测试集批量数据上评估模型，保存损失函数和准确度。我们看到，在迭代 500 次之后，测试数据集上的准确度达到 96%～97%，代码如下：

```
train_loss = []
train_acc = []
test_acc = []
for i in range(generations):
    rand_index = np.random.choice(len(train_xdata), size=batch_size)
    rand_x = train_xdata[rand_index]
    rand_x = np.expand_dims(rand_x, 3)
    rand_y = train_labels[rand_index]
    train_dict = {x_input: rand_x, y_target: rand_y}
    sess.run(train_step, feed_dict=train_dict)
    temp_train_loss, temp_train_preds = sess.run([loss, prediction], feed_dict=train_dict)
    temp_train_acc = get_accuracy(temp_train_preds, rand_y)
    if (i+1) % eval_every == 0:
        eval_index = np.random.choice(len(test_xdata), size=evaluation_size)
        eval_x = test_xdata[eval_index]
        eval_x = np.expand_dims(eval_x, 3)
        eval_y = test_labels[eval_index]
        test_dict = {eval_input: eval_x, eval_target: eval_y}
        test_preds = sess.run(test_prediction, feed_dict=test_dict)
        temp_test_acc = get_accuracy(test_preds, eval_y)
        # Record and print results
        train_loss.append(temp_train_loss)
        train_acc.append(temp_train_acc)
```

```
            test_acc.append(temp_test_acc)
            acc_and_loss = [(i+1), temp_train_loss, temp_train_acc,
temp_test_acc]
            acc_and_loss = [np.round(x,2) for x in acc_and_loss]
            print('Generation # {}. Train Loss: {:.2f}. Train Acc (Test
Acc): {:.2f} ({:.2f})'.format(*acc_and_loss))
```

13. 输出结果如下：

```
Generation # 5. Train Loss: 2.37. Train Acc (Test Acc): 7.00 (9.80)
Generation # 10. Train Loss: 2.16. Train Acc (Test Acc): 31.00
(22.00)
Generation # 15. Train Loss: 2.11. Train Acc (Test Acc): 36.00
(35.20)
...
Generation # 490. Train Loss: 0.06. Train Acc (Test Acc): 98.00
(97.40)
Generation # 495. Train Loss: 0.10. Train Acc (Test Acc): 98.00
(95.40)
Generation # 500. Train Loss: 0.14. Train Acc (Test Acc): 98.00
(96.00)
```

14. 使用 Matplotlib 模块绘制损失函数和准确度的代码，所绘图像见图 8-3：

```
eval_indices = range(0, generations, eval_every)
# Plot loss over time
plt.plot(eval_indices, train_loss, 'k-')
plt.title('Softmax Loss per Generation')
plt.xlabel('Generation')
plt.ylabel('Softmax Loss')
plt.show()

# Plot train and test accuracy
plt.plot(eval_indices, train_acc, 'k-', label='Train Set Accuracy')
plt.plot(eval_indices, test_acc, 'r--', label='Test Set Accuracy')
plt.title('Train and Test Accuracy')
plt.xlabel('Generation')
plt.ylabel('Accuracy')
plt.legend(loc='lower right')
plt.show()
```

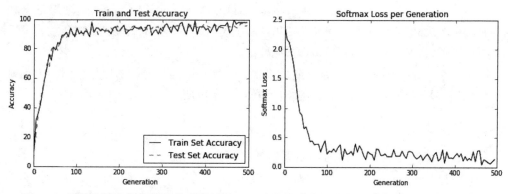

图 8-3　左图是训练集和测试集迭代训练 500 次的准确度；右图是训练集和测试集迭代训练 500 次的 softmax 损失函数值

15. 运行如下代码打印最新结果中的六幅抽样图（见图8-4）。

```
# Plot the 6 of the last batch results:
actuals = rand_y[0:6]
predictions = np.argmax(temp_train_preds,axis=1)[0:6]
images = np.squeeze(rand_x[0:6])
Nrows = 2
Ncols = 3
for i in range(6):
    plt.subplot(Nrows, Ncols, i+1)
    plt.imshow(np.reshape(images[i], [28,28]), cmap='Greys_r')
    plt.title('Actual: ' + str(actuals[i]) + ' Pred: ' +
str(predictions[i]), fontsize=10)
    frame = plt.gca()
    frame.axes.get_xaxis().set_visible(False)
    frame.axes.get_yaxis().set_visible(False)
```

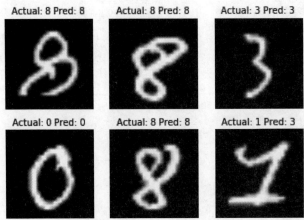

图8-4 六幅随机图，标题中是实际数字和预测数字。右下角图片预测为数字3，实际是数字1。

8.2.3 工作原理

我们提升了算法模型在MNIST数据集的性能，并且从原始数据集训练模型迅速获得了约97%的准确度。模型中前两层是卷积操作、ReLU和maxpool的组合。后两层是全联接层。在本例中，批量训练的批量大小为100，在迭代训练中观察准确度和损失函数，最后绘制六幅随机图片以及对应的实际数字和预测数字。

卷积神经网络算法在图像识别方向效果很好。部分原因是卷积层操作将图片中重要的部分特征转化成低维特征。卷积神经网络模型创建它们的特征，并用该特征预测。

8.2.4 延伸学习

在最近几年，CNN模型在图像识别领域发展迅速。有许多新观点和架构方案频出。Arxiv.org网站（https://arxiv.org/）收录该领域的很多论文，该网站是由康奈尔大学创建并维护的。Arxiv.org网站包含许多领域最新发表的文章，比如计算机科学及其子领域，如计算机视觉和图像识别（https://arxiv.org/list/cs.CV/recent）。

8.2.5 参考

下面是一些卷积神经网络算法的学习资料：

- Stanford University has a great wiki here:
  http://scarlet.stanford.edu/teach/index.php/An_Introduction_to_Convolutional_Neural_Networks
- *Deep Learning* by Michael Nielsen, found here:
  http://neuralnetworksanddeeplearning.com/chap6.html
- *An Introduction to Convolutional Neural Networks* by Jianxin Wu, found here: https://pdfs.semanticscholar.org/450c/a19932fcef1ca6d0442cbf52fec38fb9d1e5.pdf

8.3 用 TensorFlow 实现进阶的 CNN

将 CNN 模型扩展到图像识别上非常重要，我们因此可以了解如何增加网络深度。如果我们有足够大的数据集，就会提高预测的准确度。重复卷积操作、maxpool 操作和 ReLU 操作是增加神经网络深度的标准方法，许多准确度高的图像识别网络都使用该方法。

8.3.1 开始

本节将实现一个更高级的读取图像数据的方法，并使用更大的 CNN 模型进行 CIFAR10 数据集（https://www.cs.toronto.edu/~kriz/cifar.html）上的图像识别。该图片数据集有 60 000 张 32×32 像素的图片，分 10 个类别。图片可能的分类是：airplane、automobile、bird、cat、deer、dog、frog、horse、ship 和 truck，请参考 8.3.4 节中的第 1 条。

大部分图片数据集都太大，不能全部放入内存，TensorFlow 的做法是建立一个图像管道从文件中一次批量读取。为此我们会建立一个图像数据读取器，然后创建一个批量数据的队列。

一般地，对于图像识别的数据，都会在模型训练前给图片随机干扰。本例中，我们将进行随机裁剪、翻转和调节亮度。

本节是 TensorFlow 官方 CIFAR-10 例子的改写版，其中官方例子见参考部分。我们将官方例子写成脚本，并将逐行解读重要的代码。我们也会将一些常量值和参数调整为论文中引用的值，具体会在详细步骤中指明。

8.3.2 动手做

1. 导入必要的编程库，创建一个计算图会话，代码如下：

```
import os
import sys
import tarfile
import matplotlib.pyplot as plt
import numpy as np
import tensorflow as tf
from six.moves import urllib
sess = tf.Session()
```

2. 声明一些模型参数。将训练集和测试集的批量大小设为128。我们总共迭代20 000次,并且每迭代50次打印出状态值。每迭代500次,我们将在测试集的批量数据上进行模型评估。设置图片长度和宽度,以及随机裁剪图片的大小。颜色通道设为3通道(红色、绿色和蓝色),目标分类设为10类。最后声明存储数据和批量图片的位置。具体代码如下:

```
batch_size = 128
output_every = 50
generations = 20000
eval_every = 500
image_height = 32
image_width = 32
crop_height = 24
crop_width = 24
num_channels = 3
num_targets = 10
data_dir = 'temp'
extract_folder = 'cifar-10-batches-bin'
```

3. 推荐降低学习率来训练更好的模型,所以我们采用指数级减小学习率:学习率初始值设为0.1,每迭代250次指数级减少学习率,因子为10%。公式为:$0.1 \cdot 0.9^{\frac{x}{250}}$,其中 x 是当前迭代的次数。TensorFlow默认是连续减小学习率,但是也接受阶梯式更新学习率,此处设置参数以备后用:

```
learning_rate = 0.1
lr_decay = 0.9
num_gens_to_wait = 250.
```

4. 设置读取二进制CIFA-10图片的参数,代码如下:

```
image_vec_length = image_height * image_width * num_channels
record_length = 1 + image_vec_length
```

5. 设置下载CIFAR-10图像数据集的URL和数据目录,代码如下:

```
data_dir = 'temp'
if not os.path.exists(data_dir):
    os.makedirs(data_dir)
cifar10_url =
'http://www.cs.toronto.edu/~kriz/cifar-10-binary.tar.gz'
data_file = os.path.join(data_dir, 'cifar-10-binary.tar.gz')
if not os.path.isfile(data_file):
    # Download file
    filepath, _ = urllib.request.urlretrieve(cifar10_url,
data_file)
    # Extract file
    tarfile.open(filepath, 'r:gz').extractall(data_dir)
```

6. 使用read_cifar_files()函数建立图片读取器,返回一个随机打乱的图片。首先,声明一个读取固定字节长度的读取器;然后从图像队列中读取图片,抽取图片并标记;最后使用TensorFlow内建的图像修改函数随机打乱图片,代码如下:

```
def read_cifar_files(filename_queue, distort_images = True):
    reader = tf.FixedLengthRecordReader(record_bytes=record_length)
    key, record_string = reader.read(filename_queue)
```

```
        record_bytes = tf.decode_raw(record_string, tf.uint8)
        # Extract label
        image_label = tf.cast(tf.slice(record_bytes, [0], [1]), 
tf.int32)
        # Extract image
        image_extracted = tf.reshape(tf.slice(record_bytes, [1], 
[image_vec_length]), [num_channels, image_height, image_width])
        # Reshape image
        image_uint8image = tf.transpose(image_extracted, [1, 2, 0])
        reshaped_image = tf.cast(image_uint8image, tf.float32)
        # Randomly Crop image
        final_image = 
tf.image.resize_image_with_crop_or_pad(reshaped_image, crop_width, 
crop_height)
        if distort_images:
            # Randomly flip the image horizontally, change the 
brightness and contrast
            final_image = tf.image.random_flip_left_right(final_image)
            final_image = 
tf.image.random_brightness(final_image,max_delta=63)
            final_image = 
tf.image.random_contrast(final_image,lower=0.2, upper=1.8)
        # Normalize whitening
        final_image = tf.image.per_image_standardization(final_image)
        return final_image, image_label
```

7. 声明批量处理使用的图像管道填充函数。首先，需要建立读取图片的列表，定义如何用 TensorFlow 内建函数创建的 input producer 对象读取这些图片列表。把 input producer 传入上一步创建的图片读取函数 read_cifar_files() 中。然后创建图像队列的批量读取器，shuffle_batch()。具体代码如下：

```
def input_pipeline(batch_size, train_logical=True):
    if train_logical:
        files = [os.path.join(data_dir, extract_folder, 
'data_batch_{}.bin'.format(i)) for i in range(1,6)]
    else:
        files = [os.path.join(data_dir, extract_folder, 
'test_batch.bin')]
    filename_queue = tf.train.string_input_producer(files)
    image, label = read_cifar_files(filename_queue)
    min_after_dequeue = 1000
    capacity = min_after_dequeue + 3 * batch_size
    example_batch, label_batch = tf.train.shuffle_batch([image, 
label], batch_size, capacity, min_after_dequeue)
    return example_batch, label_batch
```

> 设置合适的 min_after_dequeue 值是相当重要的，该参数是设置抽样图片缓存最小值，TensorFlow 官方文档推荐设置为 (#threads + error margin)*batch_size。注意，该参数设置太大会导致更多的 shuffle。从图像队列中 shuffle 大的图像数据集需要更多的内存。

8. 声明模型函数。本例的模型使用两个卷积层，接着是三个全联接层。为了便于声明模型变量，我们将定义两个变量函数。两层卷积操作各创建64个特征。第一个全联接层联接第二个卷积层，有384个隐藏节点。第二个全联接层操作联接刚才的384个隐藏节点到192个隐藏节点。最后的隐藏层操作联接192个隐藏节点到我们预测的10个输出分类。具体见下面"#"注释部分，代码如下：

```
def cifar_cnn_model(input_images, batch_size, train_logical=True):
    def truncated_normal_var(name, shape, dtype):
        return tf.get_variable(name=name, shape=shape, dtype=dtype, initializer=tf.truncated_normal_initializer(stddev=0.05))
    def zero_var(name, shape, dtype):
        return tf.get_variable(name=name, shape=shape, dtype=dtype, initializer=tf.constant_initializer(0.0))
    # First Convolutional Layer
    with tf.variable_scope('conv1') as scope:
        # Conv_kernel is 5x5 for all 3 colors and we will create 64 features
        conv1_kernel = truncated_normal_var(name='conv_kernel1', shape=[5, 5, 3, 64], dtype=tf.float32)
        # We convolve across the image with a stride size of 1
        conv1 = tf.nn.conv2d(input_images, conv1_kernel, [1, 1, 1, 1], padding='SAME')
        # Initialize and add the bias term
        conv1_bias = zero_var(name='conv_bias1', shape=[64], dtype=tf.float32)
        conv1_add_bias = tf.nn.bias_add(conv1, conv1_bias)
        # ReLU element wise
        relu_conv1 = tf.nn.relu(conv1_add_bias)
    # Max Pooling
    pool1 = tf.nn.max_pool(relu_conv1, ksize=[1, 3, 3, 1], strides=[1, 2, 2, 1],padding='SAME', name='pool_layer1')
    # Local Response Normalization
    norm1 = tf.nn.lrn(pool1, depth_radius=5, bias=2.0, alpha=1e-3, beta=0.75, name='norm1')
    # Second Convolutional Layer
    with tf.variable_scope('conv2') as scope:
        # Conv kernel is 5x5, across all prior 64 features and we create 64 more features
        conv2_kernel = truncated_normal_var(name='conv_kernel2', shape=[5, 5, 64, 64], dtype=tf.float32)
        # Convolve filter across prior output with stride size of 1
        conv2 = tf.nn.conv2d(norm1, conv2_kernel, [1, 1, 1, 1], padding='SAME')
        # Initialize and add the bias
        conv2_bias = zero_var(name='conv_bias2', shape=[64], dtype=tf.float32)
        conv2_add_bias = tf.nn.bias_add(conv2, conv2_bias)
        # ReLU element wise
        relu_conv2 = tf.nn.relu(conv2_add_bias)
    # Max Pooling
    pool2 = tf.nn.max_pool(relu_conv2, ksize=[1, 3, 3, 1], strides=[1, 2, 2, 1], padding='SAME', name='pool_layer2')
     # Local Response Normalization (parameters from paper)
    norm2 = tf.nn.lrn(pool2, depth_radius=5, bias=2.0, alpha=1e-3, beta=0.75, name='norm2')
    # Reshape output into a single matrix for multiplication for
```

```
the fully connected layers
    reshaped_output = tf.reshape(norm2, [batch_size, -1])
    reshaped_dim = reshaped_output.get_shape()[1].value
    # First Fully Connected Layer
    with tf.variable_scope('full1') as scope:
        # Fully connected layer will have 384 outputs.
        full_weight1 = truncated_normal_var(name='full_mult1',
shape=[reshaped_dim, 384], dtype=tf.float32)
        full_bias1 = zero_var(name='full_bias1', shape=[384],
dtype=tf.float32)
        full_layer1 = tf.nn.relu(tf.add(tf.matmul(reshaped_output,
full_weight1), full_bias1))
    # Second Fully Connected Layer
    with tf.variable_scope('full2') as scope:
        # Second fully connected layer has 192 outputs.
        full_weight2 = truncated_normal_var(name='full_mult2',
shape=[384, 192], dtype=tf.float32)
        full_bias2 = zero_var(name='full_bias2', shape=[192],
dtype=tf.float32)
        full_layer2 = tf.nn.relu(tf.add(tf.matmul(full_layer1,
full_weight2), full_bias2))
    # Final Fully Connected Layer -> 10 categories for output
(num_targets)
    with tf.variable_scope('full3') as scope:
        # Final fully connected layer has 10 (num_targets) outputs.
        full_weight3 = truncated_normal_var(name='full_mult3',
shape=[192, num_targets], dtype=tf.float32)
        full_bias3 = zero_var(name='full_bias3',
shape=[num_targets], dtype=tf.float32)
        final_output = tf.add(tf.matmul(full_layer2, full_weight3),
full_bias3)
    return final_output
```

 局部响应归一化参数采用参考论文中的值，见8.3.4节。

9. 创建损失函数。本例使用 softmax 损失函数，因为一张图片应该属于其中一个类别，所以输出结果应该是10类分类的概率分布，代码如下：

```
def cifar_loss(logits, targets):
    # Get rid of extra dimensions and cast targets into integers
    targets = tf.squeeze(tf.cast(targets, tf.int32))
    # Calculate cross entropy from logits and targets
    cross_entropy =
tf.nn.sparse_softmax_cross_entropy_with_logits(logits=logits,
labels=targets)
    # Take the average loss across batch size
    cross_entropy_mean = tf.reduce_mean(cross_entropy)
    return cross_entropy_mean
```

10. 定义训练步骤函数。在训练步骤中学习率将指数级减小，代码如下：

```
def train_step(loss_value, generation_num):
    # Our learning rate is an exponential decay (stepped down)
    model_learning_rate = tf.train.exponential_decay(learning_rate,
generation_num, num_gens_to_wait, lr_decay, staircase=True)
```

```
    # Create optimizer
    my_optimizer =
tf.train.GradientDescentOptimizer(model_learning_rate)
    # Initialize train step
    train_step = my_optimizer.minimize(loss_value)
    return train_step
```

11. 创建批量图片的准确度函数。该函数输入 logits 和目标向量,输出平均准确度。训练批量图片和测试批量图片都可以使用该准确度函数,代码如下:

```
def accuracy_of_batch(logits, targets):
    # Make sure targets are integers and drop extra dimensions
    targets = tf.squeeze(tf.cast(targets, tf.int32))
    # Get predicted values by finding which logit is the greatest
    batch_predictions = tf.cast(tf.argmax(logits, 1), tf.int32)
    # Check if they are equal across the batch
    predicted_correctly = tf.equal(batch_predictions, targets)
    # Average the 1's and 0's (True's and False's) across the batch
size
    accuracy = tf.reduce_mean(tf.cast(predicted_correctly,
tf.float32))
    return accuracy
```

12. 有了图像管道函数 input_pipeline() 后,我们开始初始化训练图像管道和测试图像管道,代码如下:

```
images, targets = input_pipeline(batch_size, train_logical=True)
test_images, test_targets = input_pipeline(batch_size,
train_logical=False)
```

13. 初始化模型的训练输出和测试输出。值得注意的是,需要在创建训练模型后声明 scope.reuse_variables(),这样可以在创建测试模型时重用训练模型相同的模型参数,代码如下:

```
with tf.variable_scope('model_definition') as scope:
    # Declare the training network model
    model_output = cifar_cnn_model(images, batch_size)
    # Use same variables within scope
    scope.reuse_variables()
    # Declare test model output
    test_output = cifar_cnn_model(test_images, batch_size)
```

14. 初始化损失函数和测试准确度函数。然后声明迭代变量。该迭代变量需要声明为非训练型变量,并传入训练函数,用于计算学习率的指数级衰减值,代码如下:

```
loss = cifar_loss(model_output, targets)
accuracy = accuracy_of_batch(test_output, test_targets)
generation_num = tf.Variable(0, trainable=False)
train_op = train_step(loss, generation_num)
```

15. 初始化所有模型变量,然后运行 TensorFlow 的 start_queue_runners() 函数启动图像管道。图像管道通过赋值字典传入批量图片,开始训练模型和测试模型输出,代码如下:

```
init = tf.global_variables_initializer()
sess.run(init)
tf.train.start_queue_runners(sess=sess)
```

16. 现在遍历迭代训练,保存训练集损失函数和测试集准确度,代码如下:

```
train_loss = []
test_accuracy = []
for i in range(generations):
    _, loss_value = sess.run([train_op, loss])
    if (i+1) % output_every == 0:
        train_loss.append(loss_value)
        output = 'Generation {}: Loss = {:.5f}'.format((i+1), loss_value)
        print(output)
    if (i+1) % eval_every == 0:
        [temp_accuracy] = sess.run([accuracy])
        test_accuracy.append(temp_accuracy)
        acc_output = ' --- Test Accuracy= {:.2f}%.'.format(100. * temp_accuracy)
        print(acc_output)
```

17. 输出结果如下:

```
...
Generation 19500: Loss = 0.04461
 --- Test Accuracy = 80.47%.
Generation 19550: Loss = 0.01171
Generation 19600: Loss = 0.06911
Generation 19650: Loss = 0.08629
Generation 19700: Loss = 0.05296
Generation 19750: Loss = 0.03462
Generation 19800: Loss = 0.03182
Generation 19850: Loss = 0.07092
Generation 19900: Loss = 0.11342
Generation 19950: Loss = 0.08751
Generation 20000: Loss = 0.02228
 --- Test Accuracy = 83.59%.
```

18. 在训练中使用 matplotlib 模块绘制损失函数和准确度(见图 8-5),代码如下:

```
eval_indices = range(0, generations, eval_every)
output_indices = range(0, generations, output_every)
# Plot loss over time
plt.plot(output_indices, train_loss, 'k-')
plt.title('Softmax Loss per Generation')
plt.xlabel('Generation')
plt.ylabel('Softmax Loss')
plt.show()

# Plot accuracy over time
plt.plot(eval_indices, test_accuracy, 'k-')
plt.title('Test Accuracy')
plt.xlabel('Generation')
plt.ylabel('Accuracy')
plt.show()
```

8.3.3 工作原理

下载 CIFAR-10 图片数据集后,我们建立图像管道。关于图像管道的详细信息请见官方 TensorFlow CIFAR-10 示例。我们使用训练图片管道和测试图片管道预测图片的准确分类。

在最后,模型训练在测试数据集上达到了 75% 的准确度。

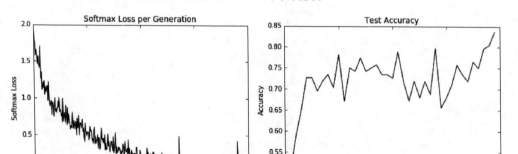

图 8-5　左图是训练集损失函数图;右图是测试集准确度图。CIFAR-10 图像识别的 CNN 模型在测试集上达到 75% 的准确度

8.3.4　参考

- For more information about the CIFAR-10 dataset, please see *Learning Multiple Layers of Features from Tiny Images*, Alex Krizhevsky, 2009: https://www.cs.toronto.edu/~kriz/learning-features-2009-TR.pdf
- To view the original TensorFlow code, please see: https://github.com/tensorflow/models/tree/master/tutorials/image/cifar10
- For more on local response normalization, please see *ImageNet Classification with Deep Convolutional Neural Networks*, Krizhevsky, A., et. al. 2012: http://papers.nips.cc/paper/4824-imagenet-classification-with-deep-convolutional-neural-networks

8.4　再训练已有的 CNN 模型

从原始数据集开始训练一个全新的图像识别模型需耗费大量时间和计算力。如果我们可以重用预训练好的网络训练图片,将会缩短计算时间。本节将展示如何使用预训练好的 TensorFlow 图像模型,微调后训练其他图片数据集。

8.4.1　开始

基本思路是重用预训练模型的卷积层的权重和结构,然后再训练全联接层。TensorFlow 官方提供一个利用已有的 CNN 模型进行再训练的例子(见 8.4.4 节)。本节将展示如何在 CIFAR-10 图片数据集上使用相同的方法。我们采用的 CNN 网络使用了非常流行的 Inception 架构。Inception CNN 模型由 Google 公司创建,并在许多图像识别基准测试中表现不俗,详情请见 8.4.4 节中的第二论文。

下面的 Python 脚本显示如何下载 CIFAR-10 图片数据集,自动分割图片数据、标注,并保存到训练集和测试集文件夹中的十个分类;然后展示如何训练图片数据集。

8.4.2 动手做

1. 导入必要的编程库，包括下载、解压和保存 CIFAR-10 图片数据的编程库，代码如下：

```
import os
import tarfile
import _pickle as cPickle
import numpy as np
import urllib.request
import scipy.misc
from imageio import imwrite
```

2. 声明 CIFAR-10 图片数据链接，创建存储数据的临时文件夹，并声明图片的十个分类，代码如下：

```
cifar_link = 
'https://www.cs.toronto.edu/~kriz/cifar-10-python.tar.gz'
data_dir = 'temp'
if not os.path.isdir(data_dir):
    os.makedirs(data_dir)
objects = ['airplane', 'automobile', 'bird', 'cat', 'deer', 'dog', 
'frog', 'horse', 'ship', 'truck']
```

3. 下载 CIFAR-10.tar 数据文件，并解压压缩文件，代码如下：

```
target_file = os.path.join(data_dir, 'cifar-10-python.tar.gz')
if not os.path.isfile(target_file):
    print('CIFAR-10 file not found. Downloading CIFAR data (Size = 163MB)')
    print('This may take a few minutes, please wait.')
    filename, headers = urllib.request.urlretrieve(cifar_link, target_file)
# Extract into memory
tar = tarfile.open(target_file)
tar.extractall(path=data_dir)
tar.close()
```

4. 创建训练所需的文件夹结构。临时目录下有两个文件夹 train_dir 和 validation_dir。每个文件夹下有 10 个子文件夹，分别存储 10 个目标分类，代码如下：

```
# Create train image folders
train_folder = 'train_dir'
if not os.path.isdir(os.path.join(data_dir, train_folder)):
    for i in range(10):
        folder = os.path.join(data_dir, train_folder, objects[i])
        os.makedirs(folder)
# Create test image folders
test_folder = 'validation_dir'
if not os.path.isdir(os.path.join(data_dir, test_folder)):
    for i in range(10):
        folder = os.path.join(data_dir, test_folder, objects[i])
        os.makedirs(folder)
```

5. 为了保存图片，我们将创建函数从内存中加载图片并存入图像字典中，代码如下：

```
def load_batch_from_file(file):
    file_conn = open(file, 'rb')
    image_dictionary = cPickle.load(file_conn, encoding='latin1')
```

```
        file_conn.close()
        return(image_dictionary)
```

6. 通过上一步的字典,将每个文件保存到正确位置,代码如下:

```
def save_images_from_dict(image_dict, folder='data_dir'):
    for ix, label in enumerate(image_dict['labels']):
        folder_path = os.path.join(data_dir, folder, objects[label])
        filename = image_dict['filenames'][ix]
        #Transform image data
        image_array = image_dict['data'][ix]
        image_array.resize([3, 32, 32])
        # Save image
        output_location = os.path.join(folder_path, filename)
        imwrite(output_location,image_array.transpose())
```

7. 对于上一步的函数,遍历下载数据文件,并把每个图片保存到正确的位置,代码如下:

```
data_location = os.path.join(data_dir, 'cifar-10-batches-py')
train_names = ['data_batch_' + str(x) for x in range(1,6)]
test_names = ['test_batch']
# Sort train images
for file in train_names:
    print('Saving images from file: {}'.format(file))
    file_location = os.path.join(data_dir, 'cifar-10-batches-py', file)
    image_dict = load_batch_from_file(file_location)
    save_images_from_dict(image_dict, folder=train_folder)
# Sort test images
for file in test_names:
    print('Saving images from file: {}'.format(file))
    file_location = os.path.join(data_dir, 'cifar-10-batches-py', file)
    image_dict = load_batch_from_file(file_location)
    save_images_from_dict(image_dict, folder=test_folder)
```

8. 程序的最后部分是创建标注文件。该文件用标注(而不是数值索引)解释输出结果,代码如下:

```
cifar_labels_file = os.path.join(data_dir,'cifar10_labels.txt')
print('Writing labels file, {}'.format(cifar_labels_file))
with open(cifar_labels_file, 'w') as labels_file:
    for item in objects:
        labels_file.write("{}n".format(item))
```

9. 上面的脚本运行之后,下载图片数据集并排序归类。接着按 TensorFlow 官方示例操作,先复制例子源码,代码如下:

```
git clone
https://github.com/tensorflow/models/tree/master/research/inception
```

10. 为了重用预训练模型,我们下载神经网络权重并应用于新神经网络模型,为此,请访问网址 https:// github. com/ tensorflow/ models/ tree/ master/ research/ slim,按照指导下载安装 cifar10 模型架构和权重。其中你将会下载一个数据目录,包括 build、train 和 test 程序脚本。

对于这个步骤，我们导航到 research/inception/inception 目录，然后执行下面的命令，使用 --train_directory、--validation_directory、--output_directory 和 --labels_file 的路径指向创建的目录结构的相对路径或完整路径。

11. 现在已经有图片在正确文件夹结构中，我们将其转为 TFRecords 对象，代码如下：

```
me@computer:~$ python3 data/build_image_data.py
--train_directory="temp/train_dir/"
--validation_directory="temp/validation_dir"
--output_directory="temp/" --labels_file="temp/cifar10_labels.txt"
```

12. 使用 bazel 模块训练算法模型，设置 fine_tune 参数为 true。该脚本每迭代 10 次输出损失函数。我们可以随时终止进程，因为模型输出结果都保存于 temp/training_results 文件夹。我们能从该文件夹加载模型数据进行模型评估，代码如下：

```
me@computer:~$ bazel-bin/inception/flowers_train

--train_dir="temp/training_results" --data_dir="temp/data_dir"
--pretrained_model_checkpoint_path="model.ckpt-157585"
--fine_tune=True --initial_learning_rate=0.001
--input_queue_memory_factor=1
```

13. 训练输出结果如下：

```
2018-06-02 11:10:10.557012: step 1290, loss = 2.02 (1.2
examples/sec; 23.771 sec/batch)
...
```

8.4.3 工作原理

TensorFlow 官方示例训练模型是基于已训练好的 CNN 模型，其要求从 CIFAR-10 图像数据创建文件夹。我们将 CIFAR-10 图像数据转化成 TFRecords 文件格式进行模型训练。注意，我们是微调已有的网络模型，重新训练全联接层来拟合 10 个目标分类。

8.4.4 参考

- Official Tensorflow Inception-v3 tutorial: https://www.tensorflow.org/tutorials/images/image_recognition
- Googlenet Inception-v3 paper: https://arxiv.org/abs/1512.00567

8.5 用 TensorFlow 实现图像风格迁移

图像识别 CNN 模型训练好之后，我们能用网络结构训练其他感兴趣的数据或者图像处理。Stylenet 程序试图学习一幅图的风格，并将该图像风格应用于另外一幅图（保持后者的图片结构或者内容）。如果我们能找到与图像风格强相关的 CNN 模型中间层节点，且节点与图像内容无关，那么就能进行风格迁移。

8.5.1 开始

Stylenet 程序需输入两幅图片,将一幅图片的图像风格应用于另外一幅图的内容上。该程序基于 2015 年发布的著名文章"A Neural Algorithm of Artistic Style",见 8.5.4 节。该文章的作者发现一些 CNN 模型的中间层存在某些属性可以编码图片风格和图片内容。最后,我们从风格图片中训练图片风格层,从原始图片中训练图片内容层,并且反向传播这些计算损失函数,从而让原始图片更像风格图片。

我们将下载文章中推荐的网络——imagenet-vgg-19。imagenet-vgg-16 网络也表现不错,但是前述文章中推荐的是 imagenet-vgg-19 网络。

8.5.2 动手做

1. 下载 .mat 格式的预训练网络,存为 .mat 文件格式。mat 文件格式是一种 matlab 对象,利用 Python 的 scipy 模块读取该文件。下面是下载 mat 对象的链接,该模型保存在 Python 脚本同一文件夹下。

http://www.vlfeat.org/matconvnet/models/beta16/imagenet-vgg-verydeep-19.mat

2. 导入必要的编程库,代码如下:

```
import os
import scipy.io
import scipy.misc
import imageio
from skimage.transform import resize
from operator import mul
from functools import reduce
import numpy as np
import tensorflow as tf
from tensorflow.python.framework import ops
ops.reset_default_graph()
```

3. 声明两幅图片(原始图片和风格图片)的位置。我们将使用本书的封面作为原始图片;梵高的大作《Starry Night》作为风格图片。这两幅图片可以在 GitHub(https://github.com/nfmcclure/tensorflow_cookbook)上下载(导航至 Stylenet 部分),代码如下:

```
original_image_file = 'temp/book_cover.jpg'
style_image_file = 'temp/starry_night.jpg'
```

4. 设置模型参数:mat 文件位置、网络权重、学习率、迭代次数和输出中间图片的频率。该权重可以增加应用于原始图片中风格图片的权重。这些超参数可以根据实际需求做出调整,代码如下:

```
vgg_path = 'imagenet-vgg-verydeep-19.mat'
original_image_weight = 5.0
style_image_weight = 500.0
regularization_weight = 100
learning_rate = 10
generations = 100
```

```
output_generations = 25
beta1 = 0.9
beta2 = 0.999
```

5. 使用 scipy 模块加载两幅图片，并将风格图片的维度调整的和原始图片一致，代码如下：

```
original_image = imageio.imread(original_image_file)
style_image = imageio.imread(style_image_file)

# Get shape of target and make the style image the same
target_shape = original_image.shape
style_image = resize(style_image, target_shape)
```

6. 从文章中获知，我们能定义各层出现的顺序，本例使用文章作者约定的名称，代码如下：

```
vgg_layers = ['conv1_1', 'relu1_1',
              'conv1_2', 'relu1_2', 'pool1',
              'conv2_1', 'relu2_1',
              'conv2_2', 'relu2_2', 'pool2',
              'conv3_1', 'relu3_1',
              'conv3_2', 'relu3_2',
'conv3_3', 'relu3_3',
'conv3_4', 'relu3_4', 'pool3',
'conv4_1', 'relu4_1',
'conv4_2', 'relu4_2',
'conv4_3', 'relu4_3',
'conv4_4', 'relu4_4', 'pool4',
'conv5_1', 'relu5_1',
'conv5_2', 'relu5_2',
'conv5_3', 'relu5_3',
'conv5_4', 'relu5_4']
```

7. 定义函数抽取 mat 文件中的参数，代码如下：

```
def extract_net_info(path_to_params):
    vgg_data = scipy.io.loadmat(path_to_params)
    normalization_matrix = vgg_data['normalization'][0][0][0]
    mat_mean = np.mean(normalization_matrix, axis=(0,1))
    network_weights = vgg_data['layers'][0]
    return mat_mean, network_weights
```

8. 基于上述加载的权重和网络层定义，通过以下函数来创建网络。迭代训练每层，并分配合适的权重和偏置，代码如下：

```
def vgg_network(network_weights, init_image):
    network = {}
    image = init_image
    for i, layer in enumerate(vgg_layers):
        if layer[1] == 'c':
            weights, bias = network_weights[i][0][0][0][0]
            weights = np.transpose(weights, (1, 0, 2, 3))
            bias = bias.reshape(-1)
            conv_layer = tf.nn.conv2d(image, tf.constant(weights),
(1, 1, 1, 1), 'SAME')
            image = tf.nn.bias_add(conv_layer, bias)
```

```
        elif layer[1] == 'r':
            image = tf.nn.relu(image)
        else:
            image = tf.nn.max_pool(image, (1, 2, 2, 1), (1, 2, 2,
1), 'SAME')
        network[layer] = image
    return(network)
```

9. 参考文章中推荐了为原始图片和风格图片分配中间层的一些策略。在本例中，原始图片采用 relu4_2 层，风格图片采用 reluX_1 层组合，代码如下：

```
original_layer = ['relu4_2']
style_layers = ['relu1_1', 'relu2_1', 'relu3_1', 'relu4_1',
'relu5_1']
```

10. 下一步运行 extract_net_info() 函数获取网络权重和平均值。我们也需要设置 VGG19 风格层权重，你可根据自己的意愿改变权重，本例中两层都设置为 0.5。

```
# Get network parameters
normalization_mean, network_weights = extract_net_info(vgg_path)
shape = (1,) + original_image.shape
style_shape = (1,) + style_image.shape
original_features = {}
style_features = {}

# Set style weights
style_weights = {l: 1./(len(style_layers)) for l in style_layers}
```

11. 为了遵从原图片的内容，需要构建内容损失函数，为此，首先需要声明 image 占位符，并创建该占位符的网络，归一化原始图片矩阵，代码如下：

```
g_original = tf.Graph()
with g_original.as_default(), tf.Session() as sess1:
    image = tf.placeholder('float', shape=shape)
    vgg_net = vgg_network(network_weights, image)
    original_minus_mean = original_image - normalization_mean
    original_norm = np.array([original_minus_mean])
    for layer in original_layers:
        original_features[layer] =
vgg_net[layer].eval(feed_dict={image: original_norm})
```

12. 如果我们想让原始图片拥有风格图片的风格，也需要构建风格损失函数，类似步骤 11，代码如下：

```
# Get style image network
g_style = tf.Graph()
with g_style.as_default(), tf.Session() as sess2:
    image = tf.placeholder('float', shape=style_shape)
    vgg_net = vgg_network(network_weights, image)
    style_minus_mean = style_image - normalization_mean
    style_norm = np.array([style_minus_mean])
    for layer in style_layers:
        features = vgg_net[layer].eval(feed_dict={image:
style_norm})
        features = np.reshape(features, (-1, features.shape[3]))
        gram = np.matmul(features.T, features) / features.size
        style_features[layer] = gram
```

13. 现在开始计算损失函数和展开训练，为此，首先初始化一张图像，带有随机噪声，作为目标图像，代码如下：

```
# Make Combined Image via loss function
with tf.Graph().as_default():
    # Get network parameters
    initial = tf.random_normal(shape) * 0.256
    init_image = tf.Variable(initial)
    vgg_net = vgg_network(network_weights, init_image)
```

14. 计算内容损失。应使得原始图像内容尽可能多的保留，代码如下：

```
# Loss from Original Image
original_layers_w = {'relu4_2': 0.5, 'relu5_2': 0.5}
original_loss = 0
for o_layer in original_layers:
    temp_original_loss = original_layers_w[o_layer] * original_image_weight *\
                (2 * tf.nn.l2_loss(vgg_net[o_layer] - original_features[o_layer]))
    original_loss += (temp_original_loss / original_features[o_layer].size)
```

15. 计算风格损失。风格损失是将风格图片的风格特征与目标图像进行对比，代码如下：

```
# Loss from Style Image
style_loss = 0
style_losses = []
for style_layer in style_layers:
    layer = vgg_net[style_layer]
    feats, height, width, channels = [x.value for x in layer.get_shape()]
    size = height * width * channels
    features = tf.reshape(layer, (-1, channels))
    style_gram_matrix = tf.matmul(tf.transpose(features), features) / size
    style_expected = style_features[style_layer]
    style_losses.append(style_weights[style_layer] * 2 *
                        tf.nn.l2_loss(style_gram_matrix - style_expected) /
                        style_expected.size)
style_loss += style_image_weight * tf.reduce_sum(style_losses)
```

16. 最后一个损失函数是总变分损失，对变化显著的邻域像素进行抑制，代码如下：

```
total_var_x = reduce(mul, init_image[:, 1:, :, :].get_shape().as_list(), 1)
total_var_y = reduce(mul, init_image[:, :, 1:, :].get_shape().as_list(), 1)
first_term = regularization_weight * 2
second_term_numerator = tf.nn.l2_loss(init_image[:, 1:, :, :] -
    init_image[:, :shape[1]-1, :, :])
second_term = second_term_numerator / total_var_y
third_term = (tf.nn.l2_loss(init_image[:, :, 1:, :] - init_image[:, :, :shape[2]-1, :]) / total_var_x)
total_variation_loss = first_term * (second_term + third_term)
```

17. 将三个损失函数合并为总损失函数，声明优化器函数和训练函数，代码如下：

```
# Combined Loss
loss = original_loss + style_loss + total_variation_loss

# Declare Optimization Algorithm
optimizer = tf.train.AdamOptimizer(learning_rate, beta1, beta2)
train_step = optimizer.minimize(loss)
```

18. 开始训练，保存中间临时图片，并保存最终输出图像，代码如下：

```
# Initialize variables and start training
with tf.Session() as sess:
    tf.global_variables_initializer().run()
    for i in range(generations):
        train_step.run()

        # Print update and save temporary output
        if (i+1) % output_generations == 0:
            print('Generation {} out of {}, loss: {}'.format(i + 1, generations, sess.run(loss)))
            image_eval = init_image.eval()
            best_image_add_mean = image_eval.reshape(shape[1:]) + normalization_mean
            output_file = 'temp_output_{}.jpg'.format(i)
            imageio.imwrite(output_file, best_image_add_mean.astype(np.uint8))
    # Save final image
    image_eval = init_image.eval()
    best_image_add_mean = image_eval.reshape(shape[1:]) + normalization_mean
    output_file = 'final_output.jpg'
    scipy.misc.imsave(output_file, best_image_add_mean)
```

图 8-6　使用风格迁移算法训练图片的 Starry Night 风格。注意，可以使用不同的权重获取不同的图片风格

8.5.3 工作原理

首先，加载两幅图片，然后加载预训练好的网络权重，为原始图片和风格图片分配网络层。我们计算三种损失函数：内容损失、风格损失和总变分损失。然后使用原始图像的内容和风格图像的风格训练随机噪声图片。

损失函数设计主要依据 GitHub 库中的风格迁移项目（https://github.com/ anishathalye/ neural-style）。为了提高性能，获得更多细节，以及得到更强的鲁棒性，建议读者阅读项目中的代码。

8.5.4 参考

- *A Neural Algorithm of Artistic Style* by Gatys, Ecker, Bethge. 2015: https://arxiv.org/abs/1508.06576
- A good recommended video of a presentation by Leon Gatys at CVPR 2016 (Computer Vision and Pattern Recognition) is here: https://www.youtube.com/watch?v=UFffxcCQMPQ .

8.6 用 TensorFlow 实现 DeepDream

重用已训练好的 CNN 模型的另外一个应用是利用一些中间节点检测标签特征的实质（比如，猫耳朵或者鸟的羽毛）。利用这个实质，我们可以找到转换任何图像的方法，以反映我们选择的任何节点的特征。本节将介绍 TensorFlow 的 DeepDream 示例，同时会详细讲解更多细节。希望通过本节的学习让读者学会使用 DeepDream 来探索 CNN 以及其中的特征。

8.6.1 开始

TensorFlow 官方示例通过一个脚本展示如何实现 DeepDream，该脚本见 8.6.4 节的第一个链接。官方例子虽好但省略了部分细节，本节将详细介绍脚本的每行代码，并微调部分代码以兼容 Python 3。

8.6.2 动手做

1. 在开始实现 DeepDream 之前，我们需要下载 GoogleNet，其为 CIFAR-1000 图片数据集上已训练好的 CNN 模型，代码如下：

```
me@computer:~$ wget
https://storage.googleapis.com/download.tensorflow.org/models/inception5h.zip
me@computer:~$ unzip inception5h.zip
```

2. 导入必要的库，并创建一个计算图会话，代码如下：

```
import os
import matplotlib.pyplot as plt
```

```
import numpy as np
import PIL.Image
import tensorflow as tf
from io import BytesIO
graph = tf.Graph()
sess = tf.InteractiveSession(graph=graph)
```

3. 声明解压的模型参数的位置,并且将这些参数加载进 TensorFlow 的计算图,代码如下:

```
# Model location
model_fn = 'tensorflow_inception_graph.pb'
# Load graph parameters
with tf.gfile.FastGFile(model_fn, 'rb') as f:
    graph_def = tf.GraphDef()
    graph_def.ParseFromString(f.read())
```

4. 创建输入数据的占位符,设置 imagenet_mean 为 117.0;然后导入计算图定义,并传入归一化的占位符,代码如下:

```
# Create placeholder for input
t_input = tf.placeholder(np.float32, name='input')
# Imagenet average bias to subtract off images
imagenet_mean = 117.0
t_preprocessed = tf.expand_dims(t_input-imagenet_mean, 0)
tf.import_graph_def(graph_def, {'input':t_preprocessed})
```

5. 导入卷积层进行可视化,并在后续处理 DeepDream 时使用,代码如下:

```
# Create a list of layers that we can refer to later
layers = [op.name for op in graph.get_operations() if
op.type=='Conv2D' and 'import/' in op.name]
# Count how many outputs for each layer
feature_nums =
[int(graph.get_tensor_by_name(name+':0').get_shape()[-1]) for name
in layers]
```

6. 现在可以选择某一层进行可视化了。我们可以通过层的名字进行选择,现在选择特征数字 139 来查看。对图片进行噪声处理,代码如下:

```
layer = 'mixed4d_3x3_bottleneck_pre_relu'
channel = 139
img_noise = np.random.uniform(size=(224,224,3)) + 100.0
```

7. 声明函数来绘制图片数组,代码如下:

```
def showarray(a, fmt='jpeg'):
    # First make sure everything is between 0 and 255
    a = np.uint8(np.clip(a, 0, 1)*255)
    # Pick an in-memory format for image display
    f = BytesIO()
    # Create the in memory image
    PIL.Image.fromarray(a).save(f, fmt)
    # Show image
    plt.imshow(a)
```

8. 在计算图中创建层迭代函数来简化重复的代码,其以层的名字来迭代,代码如下:

```
def T(layer): #Helper for getting layer output tensor return
graph.get_tensor_by_name("import/%s:0"%layer)
```

9. 下面构建一个能创建占位符的封装函数，其可以指定参数返回占位符，代码如下：

```
# The following function returns a function wrapper that will
create the placeholder
# inputs of a specified dtype
def tffunc(*argtypes):
    '''Helper that transforms TF-graph generating function into a
regular one.
    See "resize" function below.
    '''
    placeholders = list(map(tf.placeholder, argtypes))
    def wrap(f):
        out = f(*placeholders)
        def wrapper(*args, **kw):
            return out.eval(dict(zip(placeholders, args)),
session=kw.get('session'))
        return wrapper
    return wrap
```

10. 创建调整图片大小的函数，其可以指定图片大小。该函数采用 TensorFlow 的内建图片线性插值函数 tf.image.resize.bilinear()，代码如下：

```
# Helper function that uses TF to resize an image
def resize(img, size):
    img = tf.expand_dims(img, 0)
    # Change 'img' size by linear interpolation
    return tf.image.resize_bilinear(img, size)[0,:,:,:]
```

11. 现在需要一种方法更新源图片，让其更像选择的特征。我们通过指定图片的梯度如何计算来实现。我们定义函数计算图片上子区域（方格）的梯度，使得梯度计算更快。我们将在图片的 x 轴和 y 轴方向上随机移动或者滚动，这将平滑方格的影响，代码如下：

```
def calc_grad_tiled(img, t_grad, tile_size=512):
    '''Compute the value of tensor t_grad over the image in a tiled
way.
    Random shifts are applied to the image to blur tile boundaries
over
    multiple iterations.'''
    # Pick a subregion square size
    sz = tile_size
    # Get the image height and width
    h, w = img.shape[:2]
    # Get a random shift amount in the x and y direction
    sx, sy = np.random.randint(sz, size=2)
    # Randomly shift the image (roll image) in the x and y
directions
    img_shift = np.roll(np.roll(img, sx, 1), sy, 0)
    # Initialize the while image gradient as zeros
    grad = np.zeros_like(img)
    # Now we loop through all the sub-tiles in the image
    for y in range(0, max(h-sz//2, sz),sz):
        for x in range(0, max(w-sz//2, sz),sz):
            # Select the sub image tile
            sub = img_shift[y:y+sz,x:x+sz]
            # Calculate the gradient for the tile
```

```
            g = sess.run(t_grad, {t_input:sub})
            # Apply the gradient of the tile to the whole image
    gradient
            grad[y:y+sz,x:x+sz] = g
        # Return the gradient, undoing the roll operation
        return np.roll(np.roll(grad, -sx, 1), -sy, 0)
```

12. 声明 DeepDream 函数。DeepDream 算法的对象是选择特征的平均值。损失函数是基于梯度的，其依赖于输入图片和选取特征之间的距离。策略是分割图像为高频部分和低频部分，在低频部分上计算梯度。将高频部分的结果再分割为高频部分和低频部分，重复前面的过程。原始图片和低频图片称为 octaves。对传入的每个对象，计算其梯度并应用到图片中，代码如下：

```
def render_deepdream(t_obj, img0=img_noise,
                     iter_n=10, step=1.5, octave_n=4,
    octave_scale=1.4):
    # defining the optimization objective, the objective is the
    mean of the feature
    t_score = tf.reduce_mean(t_obj)
    # Our gradients will be defined as changing the t_input to get
    closer to the values of t_score. Here, t_score is the mean of the
    feature we select.
    # t_input will be the image octave (starting with the last)
    t_grad = tf.gradients(t_score, t_input)[0] # behold the power
    of automatic differentiation!
    # Store the image
    img = img0
    # Initialize the image octave list
    octaves = []
    # Since we stored the image, we need to only calculate n-1
    octaves
    for i in range(octave_n-1):
        # Extract the image shape
        hw = img.shape[:2]
        # Resize the image, scale by the octave_scale (resize by
    linear interpolation)
        lo = resize(img, np.int32(np.float32(hw)/octave_scale))
        # Residual is hi. Where residual = image - (Resize lo to
    be hw-shape)
        hi = img-resize(lo, hw)
        # Save the lo image for re-iterating
        img = lo
        # Save the extracted hi-image
        octaves.append(hi)
    # generate details octave by octave
    for octave in range(octave_n):
        if octave>0:
            # Start with the last octave
            hi = octaves[-octave]
            #
            img = resize(img, hi.shape[:2])+hi
        for i in range(iter_n):
            # Calculate gradient of the image.
            g = calc_grad_tiled(img, t_grad)
            # Ideally, we would just add the gradient, g, but
```

```
            # we want do a forward step size of it ('step'),
            # and divide it by the avg. norm of the gradient, so
            # we are adding a gradient of a certain size each step.
            # Also, to make sure we aren't dividing by zero, we add 1e-7.
            img += g*(step / (np.abs(g).mean()+1e-7))
            print('.',end = ' ')
        showarray(img/255.0)
```

13. 所有函数准备好之后，开始运行 DeepDream 算法，代码如下（对应的图见图 8-7）：

```
# Run Deep Dream
if __name__=="__main__":
    # Create resize function that has a wrapper that creates specified placeholder types
    resize = tffunc(np.float32, np.int32)(resize)
    # Open image
    img0 = PIL.Image.open('book_cover.jpg')
    img0 = np.float32(img0)
    # Show Original Image
    showarray(img0/255.0)
    # Create deep dream
    render_deepdream(T(layer)[:,:,:,139], img0, iter_n=15)
    sess.close()
```

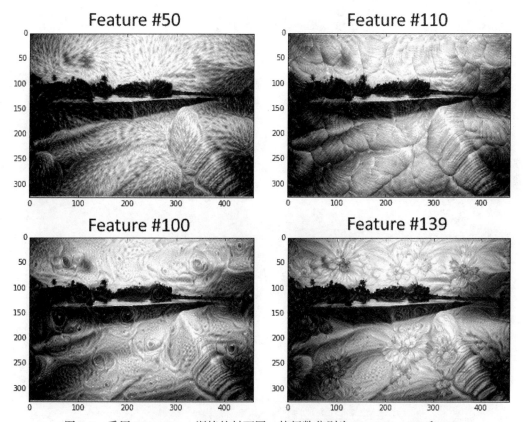

图 8-7 采用 DeepDream 训练的封面图，特征数分别为 50、110、100 和 139

8.6.3 延伸学习

我们希望读者访问官方 DeepDream 例子了解更多参考信息，也可以阅读 Google research 关于 DeepDream 的博客，见 8.6.4 节的第二个链接。

8.6.4 参考

- The TensorFlow tutorial on DeepDream:
  https://github.com/tensorflow/tensorflow/tree/master/tensorflow/examples/tutorials/deepdream
- The original Google research blog post on DeepDream:
  https://research.googleblog.com/2015/06/inceptionism-going-deeper-into-neural.html

CHAPTER 9
第 9 章

循环神经网络

本章详细介绍循环神经网络（recurrent neural network，RNN），以及如何使用 TensorFlow 实现循环神经网络。首先介绍如何使用循环神经网络预测垃圾邮件；然后基于 RNN 的变种创建莎士比亚风格文本；最后创建一个 RNN 的 Seq2Seq 翻译模型将英文翻译为德文。学完本章将掌握以下知识点：

- TensorFlow 实现 RNN 模型进行垃圾邮件预测
- TensorFlow 实现 LSTM 模型
- TensorFlow 实现堆叠多层 LSTM
- TensorFlow 实现 Seq2Seq 翻译模型
- TensorFlow 训练孪生 RNN 度量相似度

注意，本章的代码可以在 GitHub 上浏览，网址为：https://github.com/nfmcclure/tensorflow_cookbook 或 https://github.com/PacktPublishing/TensorFlow-Machine-Learning-Cookbook-Second-Edition。

9.1 简介

前面介绍的所有机器学习算法都没有考虑序列数据的情况。为了处理序列数据，我们将扩展神经网络以存储前一次迭代的输出，这类神经网络算法称为循环神经网络。全联接网络的公式为：

$$y = \sigma(Ax)$$

其中，A 为加权权重，x 为输入层。运行激励函数 σ，会返回输出层 y。如果有序列输入数据，$x_1, x_2, x_3, \cdots$，我们修改全联接层从而把之前的输入都考虑在内，表达式如下：

$$y_t = \sigma(By_{t-1} + Ax_t)$$

在循环迭代的基础上获取下一个输入。通过 softmax 函数得到概率分布输出，表达式如下：

$$s_t = softmax(Cy_t)$$

一旦有了所有序列的输出 $\{s_1, s_2, s_3, \cdots\}$，我们可以只考虑最后一个输出作为目标数值或者目标分类。图 9-1 清晰地展示了该架构是如何工作的。

我们也可以把序列输出结果作为序列（见图 9-2），即 Seq2Seq 模型：

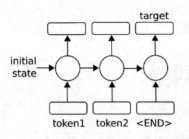

图 9-1　token 序列输入，最后一个序列输出结果作为预测输出结果（预测单个数值或者分类）

图 9-2　我们可以将序列输出结果反馈回模型生成多个输出结果（以预测一个序列）

对于任意长度的序列，利用反向传播算法训练创建长时间依赖的梯度。这样会出现梯度消失或者梯度爆炸的问题。在本章后续部分，我们将探索 RNN 单元的扩展——LSTM（Long Short Term Memory，长短期记忆）单元以解决该问题。基本思想是 LSTM 单元引入门操作（gate），该门操作控制序列上信息的流动。后续章节将会详细讲解。

 当 RNN 模型处理 NLP 时，编码用来描述将数据（NLP 中的单词或者字符）转换为数值型 RNN 特征的过程；解码用来描述将数值型 RNN 特征转换成输出的单词或者字符的过程。

9.2　用 TensorFlow 实现 RNN 模型进行垃圾邮件预测

应用标准的 RNN 单元预测奇异数值型输出，该输出是垃圾邮件的概率。

9.2.1　开始

本节将用 TensorFlow 实现一个标准的 RNN 模型，预测文本邮件是正常邮件还是垃圾邮件。本例使用 UCI 大学的机器学习仓库中的 SMS 垃圾邮件数据集。本例使用的预测架构是，嵌入文本中的输入 RNN 序列，取最后一个 RNN 输出作为是否为垃圾邮件（1 或 0）的预测。

9.2.2　动手做

1. 导入必要的编程库，代码如下：

```
import os
import re
import io
import requests
```

```
import numpy as np
import matplotlib.pyplot as plt
import tensorflow as tf
from zipfile import ZipFile
```

2. 开始计算图会话,并设置 RNN 模型参数。训练数据 20 个 epoch,批量大小为 250。邮件最大长度为 25 个单词,超过的部分会被截取掉,不够的部分用 0 填充。RNN 模型由 10 个单元组成。我们仅仅处理词频超过 10 的单词,每个单词会嵌入在长度为 50 的词向量中。dropout 概率为占位符,训练模型时设为 0.5,评估模型时设为 1.0。具体代码如下:

```
sess = tf.Session()
epochs = 20
batch_size = 250
max_sequence_length = 25
rnn_size = 10
embedding_size = 50
min_word_frequency = 10
learning_rate = 0.0005
dropout_keep_prob = tf.placeholder(tf.float32)
```

3. 获取 SMS 文本数据集。首先,在下载文本数据集前检查是否已下载过。如果已经下载过数据集,直接从文件中读取,代码如下:

```
data_dir = 'temp'
data_file = 'text_data.txt'
if not os.path.exists(data_dir):
    os.makedirs(data_dir)
if not os.path.isfile(os.path.join(data_dir, data_file)):
    zip_url = 'http://archive.ics.uci.edu/ml/machine-learning-databases/00228/sms spamcollection.zip'
    r = requests.get(zip_url)
    z = ZipFile(io.BytesIO(r.content))
    file = z.read('SMSSpamCollection')
    # Format Data
    text_data = file.decode()
    text_data = text_data.encode('ascii',errors='ignore')
    text_data = text_data.decode().split('\n')
    # Save data to text file
    with open(os.path.join(data_dir, data_file), 'w') as file_conn:
        for text in text_data:
            file_conn.write("{}\n".format(text))
else:
    # Open data from text file
    text_data = []
    with open(os.path.join(data_dir, data_file), 'r') as file_conn:
        for row in file_conn:
            text_data.append(row)
    text_data = text_data[:-1]
text_data = [x.split('\t') for x in text_data if len(x)>=1]
[text_data_target, text_data_train] = [list(x) for x in zip(*text_data)]
```

4. 为降低词汇量,我们将清洗文本数据集,移除特殊字符和空格,将所有文本转为小写,代码如下:

```
def clean_text(text_string):
    text_string = re.sub(r'([^sw]|_|[0-9])+', '', text_string)
    text_string = " ".join(text_string.split())
    text_string = text_string.lower()
    return text_string

# Clean texts
text_data_train = [clean_text(x) for x in text_data_train]
```

 注意从文本数据中移除特殊字符的清洗步骤,有时可以用空格替换该特殊字符。在理想情况下,需要根据数据集的格式选择具体的方法处理。

5. 使用 TensorFlow 内建的词汇处理器处理文本。该步骤将文本转换为索引列表,代码如下:

```
vocab_processor =
tf.contrib.learn.preprocessing.VocabularyProcessor(max_sequence_len
gth,    min_frequency=min_word_frequency)
text_processed =
np.array(list(vocab_processor.fit_transform(text_data_train)))
```

注意 tensorflow1.10 并不支持 contrib.learn.preprocessing 中的函数,一个替代建议是使用 TensorFlow 预处理包,但是目前只支持 Python2。将 TensorFlow 预处理包扩展到 Python3 的方法正在开发,并将会替代代码中的前两行。记住当前代码和升级代码在以下 GitHub 库可以获取:https://www.github.com/nfmcclure/tensorflow_cookbook 或 https://github.com/PacktPublishing/TensorFlow-Machine-Learning-Cookbook-Second-Edition。

6. 随机 shuffle 文本数据集,代码如下:

```
text_processed = np.array(text_processed)
text_data_target = np.array([1 if x=='ham' else 0 for x in
text_data_target])
shuffled_ix =
np.random.permutation(np.arange(len(text_data_target)))
x_shuffled = text_processed[shuffled_ix]
y_shuffled = text_data_target[shuffled_ix]
```

7. 分割数据集为 80-20 的训练 – 测试数据集,代码如下:

```
ix_cutoff = int(len(y_shuffled)*0.80)
x_train, x_test = x_shuffled[:ix_cutoff], x_shuffled[ix_cutoff:]
y_train, y_test = y_shuffled[:ix_cutoff], y_shuffled[ix_cutoff:]
vocab_size = len(vocab_processor.vocabulary_)
print("Vocabulary Size: {:d}".format(vocab_size))
print("80-20 Train Test split: {:d} -- {:d}".format(len(y_train),
len(y_test)))
```

 本小节我们不准备做超参数调优。如果读者有这方面的需求,请在预处理前将数据集分割为训练集 – 测试集 – 验证集。scikit-learn 的 model_selection.train_test_split() 函数可以随机分割(划分)训练集和测试集。

8. 声明计算图的占位符。输入数据 x_data 是形状为 [None, max_sequence_length] 的占位符，其以文本最大允许的长度为批量大小。输出结果 y_output 的占位符为整数 0 或者 1，代表正常邮件或者垃圾邮件，代码如下：

```
x_data = tf.placeholder(tf.int32, [None, max_sequence_length])
y_output = tf.placeholder(tf.int32, [None])
```

9. 创建输入数据 x_data 的嵌入矩阵和嵌入查找操作，代码如下：

```
embedding_mat = tf.Variable(tf.random_uniform([vocab_size, embedding_size], -1.0, 1.0))
embedding_output = tf.nn.embedding_lookup(embedding_mat, x_data)
```

10. 声明算法模型。首先，初始化 RNN 单元的类型，大小为 10。然后通过动态 RNN 函数 tf.nn.dynamic_rnn() 创建 RNN 序列，接着增加 dropout 操作，代码如下：

```
cell = tf.nn.rnn_cell.BasicRNNCell(num_units = rnn_size)
output, state = tf.nn.dynamic_rnn(cell, embedding_output, dtype=tf.float32)
output = tf.nn.dropout(output, dropout_keep_prob)
```

注意，动态 RNN 允许变长序列。即使本例所使用的是固定长度的序列，我们也推荐使用 TensorFlow 的 tf.nn.dynamic_rnn() 函数。主要原因是：实践证明动态 RNN 实际计算更快，并且允许 RNN 中运行不同长度的序列。

11. 为了进行预测，转置并重新排列 RNN 的输出结果，剪切最后的输出结果，代码如下：

```
output = tf.transpose(output, [1, 0, 2])
last = tf.gather(output, int(output.get_shape()[0]) - 1)
```

12. 为了完成 RNN 预测，我们通过全联接层将 rnn_size 大小的输出转换为二分类输出，代码如下：

```
weight = tf.Variable(tf.truncated_normal([rnn_size, 2], stddev=0.1))
bias = tf.Variable(tf.constant(0.1, shape=[2]))
logits_out = tf.nn.softmax(tf.matmul(last, weight) + bias)
```

13. 声明损失函数。本例使用 TensorFlow 的 sparse_softmax 函数，目标值是 int 型索引，logits 是 float 型，代码如下：

```
losses = tf.nn.sparse_softmax_cross_entropy_with_logits(logits=logits_out, labels=y_output)
loss = tf.reduce_mean(losses)
```

14. 创建准确度函数，比较训练集和测试集的结果，代码如下：

```
accuracy = tf.reduce_mean(tf.cast(tf.equal(tf.argmax(logits_out, 1), tf.cast(y_output, tf.int64)), tf.float32))
```

15. 创建优化器函数，初始化模型变量，代码如下：

```
optimizer = tf.train.RMSPropOptimizer(learning_rate)
train_step = optimizer.minimize(loss)
init = tf.global_variables_initializer()
sess.run(init)
```

16. 开始遍历迭代训练模型。遍历数据集多次，最佳实践表明：为了避免过拟合，每个 epoch 都需随机 shuffle 数据，代码如下：

```
train_loss = []
test_loss = []
train_accuracy = []
test_accuracy = []
# Start training
for epoch in range(epochs):
    # Shuffle training data
    shuffled_ix = np.random.permutation(np.arange(len(x_train)))
    x_train = x_train[shuffled_ix]
    y_train = y_train[shuffled_ix]
    num_batches = int(len(x_train)/batch_size) + 1
    for i in range(num_batches):
        # Select train data
        min_ix = i * batch_size
        max_ix = np.min([len(x_train), ((i+1) * batch_size)])
        x_train_batch = x_train[min_ix:max_ix]
        y_train_batch = y_train[min_ix:max_ix]
        # Run train step
        train_dict = {x_data: x_train_batch, y_output: y_train_batch, dropout_keep_prob:0.5}
        sess.run(train_step, feed_dict=train_dict)
    # Run loss and accuracy for training
    temp_train_loss, temp_train_acc = sess.run([loss, accuracy], feed_dict=train_dict)
    train_loss.append(temp_train_loss)
    train_accuracy.append(temp_train_acc)
    # Run Eval Step
    test_dict = {x_data: x_test, y_output: y_test, dropout_keep_prob:1.0}
    temp_test_loss, temp_test_acc = sess.run([loss, accuracy], feed_dict=test_dict)
    test_loss.append(temp_test_loss)
    test_accuracy.append(temp_test_acc)
    print('Epoch: {}, Test Loss: {:.2}, Test Acc: {:.2}'.format(epoch+1, temp_test_loss, temp_test_acc))
```

17. 输出结果如下：

```
Vocabulary Size: 933
80-20 Train Test split: 4459 -- 1115
Epoch: 1, Test Loss: 0.59, Test Acc: 0.83
Epoch: 2, Test Loss: 0.58, Test Acc: 0.83
...
Epoch: 19, Test Loss: 0.46, Test Acc: 0.86
Epoch: 20, Test Loss: 0.46, Test Acc: 0.86
```

18. 绘制训练集、测试集损失和准确度的代码如下：

```
epoch_seq = np.arange(1, epochs+1)
plt.plot(epoch_seq, train_loss, 'k--', label='Train Set')
```

```
plt.plot(epoch_seq, test_loss, 'r-', label='Test Set')
plt.title('Softmax Loss')
plt.xlabel('Epochs')
plt.ylabel('Softmax Loss')
plt.legend(loc='upper left')
plt.show()
# Plot accuracy over time
plt.plot(epoch_seq, train_accuracy, 'k--', label='Train Set')
plt.plot(epoch_seq, test_accuracy, 'r-', label='Test Set')
plt.title('Test Accuracy')
plt.xlabel('Epochs')
plt.ylabel('Accuracy')
plt.legend(loc='upper left')
plt.show()
```

9.2.3 工作原理

在本节中，我们创建 RNN 分类模型预测 SMS 邮件文本是否为垃圾邮件。本例在测试集上的训练准确度为 86%。图 9-3 是测试集和训练集的准确度和损失函数图。

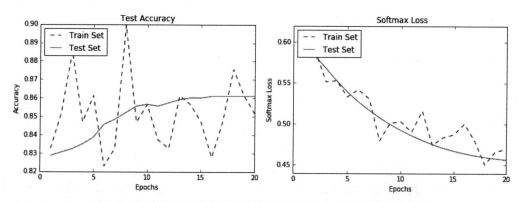

图 9-3 训练集和测试集的准确度图（左图）和损失函数图（右图）

9.2.4 延伸学习

对于序列数据，强烈推荐对训练集（全部数据）进行多次训练（对于非序列数据也推荐这样处理）。全部数据进行一次训练称为 epoch。同时强烈建议，在每个 epoch 之前进行随机 shuffle 数据。

9.3 用 TensorFlow 实现 LSTM 模型

本节通过引入 LSTM 单元，将 RNN 模型扩展为可以处理长序列的模型。

9.3.1 开始

LSTM 是传统循环神经网络的变种。LSTM 可以解决变长 RNN 模型的梯度消失或者爆炸的问题。为了解决梯度问题，LSTM 单元引入一个内部忘记门（forget gate），该门操作可以修改一个单元到下一个单元的信息流转。为了清晰阐明 LSTM 是如何工作的，我们将逐步分析 LSTM 公式的无偏版本。第一步，与常规的 RNN 模型一样：

$$i_t = \sigma\,(B_i h_{t-1} + A_i x_t)$$

为了指明什么信息可以遗忘或者通过，我们通过下面的公式评估候选信息。这些信息称为记忆单元：

$$C_t = \tanh\,(B_C h_{t-1} + A_C x_t)$$

现在我们通过忘记矩阵（forget matrix）修改可选的记忆单元，其计算公式如下：

$$f_t = \sigma\,(B_f h_{t-1} + A_f x_t)$$

将前面的记忆信息与遗忘记忆结合，然后与可选的记忆单元相加得到新的记忆信息：

$$N_t = i_t \cdot C_t + f_t N_{t-1}$$

结合前面的所有项获得记忆单元的输出：

$$O_t = \sigma\,(B_O h_{t-1} + A_O x_t + D_O N_t)$$

之后，通过迭代更新 h，计算公式：

$$h_t = O_t \cdot \tanh\,(N_t)$$

LSTM 的理念是通过记忆单元忘记或者修改输入信息。

> 本例使用 TensorFlow 的一大好处是，我们不必维护这些操作和相关的反向传播状态。TensorFlow 会维护这些状态信息，并根据模型的损失函数、优化器函数和学习率计算梯度来自动更新模型变量值。

本节将会使用带有 LSTM 结构的序列 RNN 模型在莎士比亚文本数据集上训练，预测下一个单词。我们将为该模型传入短语（比如，*thou art more*），看训练的模型是否可以预测出短语接下来的单词。

9.3.2 动手做

1. 导入必要的编程库，代码如下：

```
import os
import re
import string
import requests
import numpy as np
import collections
import random
import pickle
import matplotlib.pyplot as plt
import tensorflow as tf
```

2. 开始计算图会话，并设置 RNN 参数，代码如下：

```
sess = tf.Session()

# Set RNN Parameters
min_word_freq = 5
rnn_size = 128
epochs = 10
batch_size = 100
learning_rate = 0.001
training_seq_len = 50
embedding_size = rnn_size
save_every = 500
eval_every = 50
prime_texts = ['thou art more', 'to be or not to', 'wherefore art
thou']
```

3. 定义数据和模型的文件夹和文件名。我们将保留连字符和省略符，因为莎士比亚频繁地使用这些字符来组合单词和音节，代码如下：

```
data_dir = 'temp'
data_file = 'shakespeare.txt'
model_path = 'shakespeare_model'
full_model_dir = os.path.join(data_dir, model_path)
# Declare punctuation to remove, everything except hyphens and
apostrophe's
punctuation = string.punctuation
punctuation = ''.join([x for x in punctuation if x not in ['-',
"'"]])
```

4. 下载文本数据集。如果该数据集存在，我们将直接加载数据；如果不存在，我们将下载该文本数据集，并保存，代码如下：

```
if not os.path.exists(full_model_dir):
    os.makedirs(full_model_dir)
# Make data directory
if not os.path.exists(data_dir):
    os.makedirs(data_dir)
print('Loading Shakespeare Data')
# Check if file is downloaded.
if not os.path.isfile(os.path.join(data_dir, data_file)):
    print('Not found, downloading Shakespeare texts from
www.gutenberg.org')
    shakespeare_url =
'http://www.gutenberg.org/cache/epub/100/pg100.txt'
    # Get Shakespeare text
    response = requests.get(shakespeare_url)
    shakespeare_file = response.content
    # Decode binary into string
    s_text = shakespeare_file.decode('utf-8')
    # Drop first few descriptive paragraphs.
    s_text = s_text[7675:]
    # Remove newlines
    s_text = s_text.replace('\r\n', '')
    s_text = s_text.replace('\n', '')
    # Write to file
    with open(os.path.join(data_dir, data_file), 'w') as out_conn:
        out_conn.write(s_text)
else:
```

```
# If file has been saved, load from that file
with open(os.path.join(data_dir, data_file), 'r') as file_conn:
    s_text = file_conn.read().replace('\n', '')
```

5. 清洗莎士比亚文本，移除标点符号和多余的空格，代码如下：

```
s_text = re.sub(r'[{}]'.format(punctuation), ' ', s_text)
s_text = re.sub('\s+', ' ', s_text ).strip().lower()
```

6. 创建莎士比亚词汇表。我们创建 build_vocab() 返回两个单词字典（单词到索引的映射和索引到单词的映射），其中出现的单词要符合频次要求，代码如下：

```
def build_vocab(text, min_word_freq):
    word_counts = collections.Counter(text.split(' '))
    # limit word counts to those more frequent than cutoff
    word_counts = {key:val for key, val in word_counts.items() if val>min_word_freq}
    # Create vocab --> index mapping
    words = word_counts.keys()
    vocab_to_ix_dict = {key:(ix+1) for ix, key in enumerate(words)}
    # Add unknown key --> 0 index
    vocab_to_ix_dict['unknown']=0
    # Create index --> vocab mapping
    ix_to_vocab_dict = {val:key for key,val in vocab_to_ix_dict.items()}
    return ix_to_vocab_dict, vocab_to_ix_dict
ix2vocab, vocab2ix = build_vocab(s_text, min_word_freq)
vocab_size = len(ix2vocab) + 1
```

处理文本时，我们需要注意单词索引为 0 的值，将其保存并填充。对于未知单词也采取相同方法处理。

7. 有了单词词汇表，我们将莎士比亚文本转换成索引数组，代码如下：

```
s_text_words = s_text.split(' ')
s_text_ix = []
for ix, x in enumerate(s_text_words):
    try:
        s_text_ix.append(vocab2ix[x])
    except:
        s_text_ix.append(0)
s_text_ix = np.array(s_text_ix)
```

8. 本例将展示如何用 class 对象创建算法模型。我们将使用相同的模型（相同模型参数）来训练批量数据和抽样生成的文本。如果没有 class 对象，将很难用抽样方法训练相同的模型。在理想情况下，该 class 代码单独保存在一个 Python 文件中，它可以在脚本起始位置导入，代码如下：

```
class LSTM_Model():
    def __init__(self, rnn_size, batch_size, learning_rate,
                 training_seq_len, vocab_size, infer =False):
        self.rnn_size = rnn_size
        self.vocab_size = vocab_size
```

```python
            self.infer = infer
            self.learning_rate = learning_rate
            if infer:
                self.batch_size = 1
                self.training_seq_len = 1
            else:
                self.batch_size = batch_size
                self.training_seq_len = training_seq_len
            self.lstm_cell = tf.nn.rnn_cell.BasicLSTMCell(rnn_size)
            self.initial_state = self.lstm_cell.zero_state(self.batch_size, tf.float32)
            self.x_data = tf.placeholder(tf.int32, [self.batch_size, self.training_seq_len])
            self.y_output = tf.placeholder(tf.int32, [self.batch_size, self.training_seq_len])
            with tf.variable_scope('lstm_vars'):
                # Softmax Output Weights
                W = tf.get_variable('W', [self.rnn_size, self.vocab_size], tf.float32, tf.random_normal_initializer())
                b = tf.get_variable('b', [self.vocab_size], tf.float32, tf.constant_initializer(0.0))
                # Define Embedding
                embedding_mat = tf.get_variable('embedding_mat', [self.vocab_size, self.rnn_size], tf.float32, tf.random_normal_initializer())
                embedding_output = tf.nn.embedding_lookup(embedding_mat, self.x_data)
                rnn_inputs = tf.split(embedding_output, num_or_size_splits=self.training_seq_len, axis=1)
                rnn_inputs_trimmed = [tf.squeeze(x, [1]) for x in rnn_inputs]
            # If we are inferring (generating text), we add a 'loop' function
            # Define how to get the i+1 th input from the i th output
            def inferred_loop(prev, count):
                prev_transformed = tf.matmul(prev, W) + b
                prev_symbol = tf.stop_gradient(tf.argmax(prev_transformed, 1))
                output = tf.nn.embedding_lookup(embedding_mat, prev_symbol)
                return output
            decoder = tf.nn.seq2seq.rnn_decoder
            outputs, last_state = decoder(rnn_inputs_trimmed,
                                          self.initial_state,
                                          self.lstm_cell,
                                          loop_function=inferred_loop if infer else None)
            # Non inferred outputs
            output = tf.reshape(tf.concat(1, outputs), [-1, self.rnn_size])
            # Logits and output
            self.logit_output = tf.matmul(output, W) + b
            self.model_output = tf.nn.softmax(self.logit_output)
            loss_fun = tf.contrib.legacy_seq2seq.sequence_loss_by_example
            loss = loss_fun([self.logit_output],[tf.reshape(self.y_output, [-1])],
```

```python
                [tf.ones([self.batch_size *
self.training_seq_len])],
                  self.vocab_size)
        self.cost = tf.reduce_sum(loss) / (self.batch_size *
self.training_seq_len)
        self.final_state = last_state
        gradients, _ =
tf.clip_by_global_norm(tf.gradients(self.cost,
tf.trainable_variables()), 4.5)
        optimizer = tf.train.AdamOptimizer(self.learning_rate)
        self.train_op = optimizer.apply_gradients(zip(gradients,
tf.trainable_variables()))
    def sample(self, sess, words=ix2vocab, vocab=vocab2ix, num=10,
prime_text='thou art'):
        state = sess.run(self.lstm_cell.zero_state(1, tf.float32))
        word_list = prime_text.split()
        for word in word_list[:-1]:
            x = np.zeros((1, 1))
            x[0, 0] = vocab[word]
            feed_dict = {self.x_data: x, self.initial_state:state}
            [state] = sess.run([self.final_state],
feed_dict=feed_dict)
        out_sentence = prime_text
        word = word_list[-1]
        for n in range(num):
            x = np.zeros((1, 1))
            x[0, 0] = vocab[word]
            feed_dict = {self.x_data: x, self.initial_state:state}
            [model_output, state] = sess.run([self.model_output,
self.final_state], feed_dict=feed_dict)
            sample = np.argmax(model_output[0])
            if sample == 0:
                break
            word = words[sample]
            out_sentence = out_sentence + ' ' + word
        return out_sentence
```

9. 声明LSTM模型及其测试模型。使用tf.variable_scope管理模型变量，使得测试LSTM模型可以复用训练LSTM模型相同的参数，代码如下：

```python
with tf.variable_scope('lstm_model', reuse=tf.AUTO_REUSE) as scope:
    # Define LSTM Model
    lstm_model = LSTM_Model(rnn_size, batch_size, learning_rate,
                 training_seq_len, vocab_size)
    scope.reuse_variables()
    test_lstm_model = LSTM_Model(rnn_size, batch_size,
learning_rate,
                 training_seq_len, vocab_size, infer=True)
```

10. 创建saver操作，并分割输入文本为相同批量大小的块，然后初始化模型变量，代码如下：

```python
saver = tf.train.Saver()
# Create batches for each epoch
num_batches = int(len(s_text_ix)/(batch_size * training_seq_len)) +
1
# Split up text indices into subarrays, of equal size
```

```
batches = np.array_split(s_text_ix, num_batches)
# Reshape each split into [batch_size, training_seq_len]
batches = [np.resize(x, [batch_size, training_seq_len]) for x in
batches]
# Initialize all variables
init = tf.global_variables_initializer()
sess.run(init)
```

11. 现在通过 epoch 迭代训练，并在每个 epoch 之前将数据 shuffle。虽然文本数据是相同的，但是会用 numpy.roll() 函数改变顺序，代码如下：

```
train_loss = []
iteration_count = 1
for epoch in range(epochs):
    # Shuffle word indices
    random.shuffle(batches)
    # Create targets from shuffled batches
    targets = [np.roll(x, -1, axis=1) for x in batches]
    # Run a through one epoch
    print('Starting Epoch #{} of {}.'.format(epoch+1, epochs))
    # Reset initial LSTM state every epoch
    state = sess.run(lstm_model.initial_state)
    for ix, batch in enumerate(batches):
        training_dict = {lstm_model.x_data: batch,
lstm_model.y_output: targets[ix]}
        c, h = lstm_model.initial_state
        training_dict[c] = state.c
        training_dict[h] = state.h
        temp_loss, state, _ = sess.run([lstm_model.cost,
lstm_model.final_state, lstm_model.train_op],
feed_dict=training_dict)
        train_loss.append(temp_loss)
        # Print status every 10 gens
        if iteration_count % 10 == 0:
            summary_nums = (iteration_count, epoch+1, ix+1,
num_batches+1, temp_loss)
            print('Iteration: {}, Epoch: {}, Batch: {} out of {},
Loss: {:.2f}'.format(*summary_nums))
        # Save the model and the vocab
        if iteration_count % save_every == 0:
            # Save model
            model_file_name = os.path.join(full_model_dir, 'model')
            saver.save(sess, model_file_name, global_step =
iteration_count)
            print('Model Saved To: {}'.format(model_file_name))
            # Save vocabulary
            dictionary_file = os.path.join(full_model_dir,
'vocab.pkl')
            with open(dictionary_file, 'wb') as dict_file_conn:
                pickle.dump([vocab2ix, ix2vocab], dict_file_conn)
        if iteration_count % eval_every == 0:
            for sample in prime_texts:
                print(test_lstm_model.sample(sess, ix2vocab,
vocab2ix, num=10, prime_text=sample))
        iteration_count += 1
```

12. 输出结果如下：

```
Loading Shakespeare Data
Cleaning Text
Building Shakespeare Vocab
Vocabulary Length = 8009
Starting Epoch #1 of 10.
Iteration: 10, Epoch: 1, Batch: 10 out of 182, Loss: 10.37
Iteration: 20, Epoch: 1, Batch: 20 out of 182, Loss: 9.54
...
Iteration: 1790, Epoch: 10, Batch: 161 out of 182, Loss: 5.68
Iteration: 1800, Epoch: 10, Batch: 171 out of 182, Loss: 6.05
thou art more than i am a
to be or not to the man i have
wherefore art thou art of the long
Iteration: 1810, Epoch: 10, Batch: 181 out of 182, Loss: 5.99
```

13. 绘制训练损失随 epoch 的趋势图（见图9-4），代码如下：

```
plt.plot(train_loss, 'k-')
plt.title('Sequence to Sequence Loss')
plt.xlabel('Generation')
plt.ylabel('Loss')
plt.show()
```

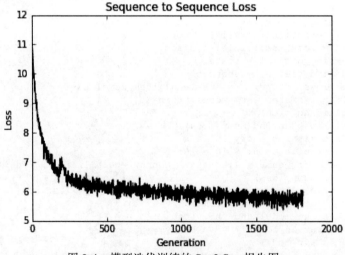

图 9-4　模型迭代训练的 Seq2 Seq 损失图

9.3.3　工作原理

在本例中，我们基于莎士比亚词汇构建带有 LSTM 结构的 RNN 模型预测下一个单词。通过增加序列大小、降低学习率，或者增加模型的 epoch，可能会提高该模型的预测效果。

9.3.4　延伸学习

对于抽样的方法，我们实现的是贪婪抽样法。贪婪抽样法会不断地出现重复的短语。例如，可能重复地说"for the for the for the…."为了防止出现该问题，我们也可以实现一个随机抽样单词的方法，通过基于输出结果的 logits 或者概率分布进行加权抽样。

9.4 TensorFlow 堆叠多层 LSTM

如同增加神经网络或者 CNN 的深度一样，我们也可以增加 RNN 的深度。本节应用三层 LSTM 促进莎士比亚文本的生成。

9.4.1 开始

我们通过堆叠多个 LSTM 层增加 RNN 模型的深度。必要时，我们可以将目标输出作为输入赋值给另外一个网络。为了看清其中的工作原理，这里以两层网络举例，见图 9-5。

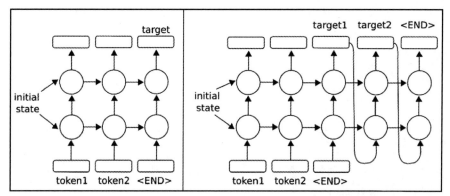

图 9-5 将一层 RNN 扩展为两层 RNN。左图是图 9-1 的扩展，通过序列输入预测一个输出；右图是图 9-2 的扩展，将输出反馈至输入，预测一个序列

TensorFlow 使用 MultiRNNCell() 函数轻松地实现多层组合，该函数输入参数为 RNN 单元列表。用 Python 调用 MultiRNNCell([rnn_cell]*num_layers) 很容易创建一个多层 RNN。

本节将展示前一节中相同的莎士比亚文本的预测。但是本例有两个变化：第一个变化是采用堆叠三层 LSTM 代替原有的一层网络；第二个变化是字符级别的预测代替单词级别的预测。字符级别的预测将极大地减少词汇表，表中仅有 40 个字符（26 个字母，10 个数字，1 个空格，3 个特殊字符）。

9.4.2 动手做

为了不重复展示和上一节相同的代码，本例只解释有区别的代码。完整源代码见 GitHub，地址为：https://github.com/nfmcclure/tensorflow_cookbook 或 https://github.com/ PacktPublishing/TensorFlow-Machine-Learning-Cookbook- Second-Edition。

1. 设置 RNN 模型的层数。该参数放在脚本起始位置，其他模型参数如下：

```
num_layers = 3
min_word_freq = 5
rnn_size = 128
epochs = 10
```

2. 第一个主要的变化是，我们将以字符来加载、处理和传入文本，而不是单词。为此，在清洗完文本数据之后，通过 Python 的 list() 函数分割整个文本，代码如下：

```
s_text = re.sub(r'[{}]'.format(punctuation), ' ', s_text)
s_text = re.sub('s+', ' ', s_text).strip().lower()
# Split up by characters
char_list = list(s_text)
```

3. 现在需要改变原有一层 LSTM 模型为多层。接收 num_layers 变量，然后利用 TensorFlow 的 MultiRNNCell() 函数创建多层 RNN 模型，代码如下：

```
class LSTM_Model():
    def __init__(self, rnn_size, num_layers, batch_size, learning_rate,
                 training_seq_len, vocab_size, infer_sample=False):
        self.rnn_size = rnn_size
        self.num_layers = num_layers
        self.vocab_size = vocab_size
        self.infer_sample = infer_sample
        self.learning_rate = learning_rate
        ...
        self.lstm_cell = tf.contrib.rnn.BasicLSTMCell(rnn_size)
        self.lstm_cell = tf.contrib.rnn.MultiRNNCell([self.lstm_cell for _ in range(self.num_layers)])
        self.initial_state = self.lstm_cell.zero_state(self.batch_size, tf.float32)
        self.x_data = tf.placeholder(tf.int32, [self.batch_size, self.training_seq_len])
        self.y_output = tf.placeholder(tf.int32, [self.batch_size, self.training_seq_len])
```

TensorFlow 的 MultiRNNCell() 函数的输入参数为 RNN 单元列表。本例中的 RNN 层是相同的，但是你也可以采用任何 RNN 层堆叠组合。

4. 其他代码都与上一节相同，这里不再赘述。训练模型输出如下：

```
Building Shakespeare Vocab by Characters
Vocabulary Length = 40
Starting Epoch #1 of 10
Iteration: 9430, Epoch: 10, Batch: 889 out of 950, Loss: 1.54
Iteration: 9440, Epoch: 10, Batch: 899 out of 950, Loss: 1.46
Iteration: 9450, Epoch: 10, Batch: 909 out of 950, Loss: 1.49
thou art more than the
to be or not to the serva
wherefore art thou dost thou
Iteration: 9460, Epoch: 10, Batch: 919 out of 950, Loss: 1.41
Iteration: 9470, Epoch: 10, Batch: 929 out of 950, Loss: 1.45
Iteration: 9480, Epoch: 10, Batch: 939 out of 950, Loss: 1.59
Iteration: 9490, Epoch: 10, Batch: 949 out of 950, Loss: 1.42
```

5. 最后的输出文本的抽样如下：

```
thou art more fancy with to be or not to be for be wherefore art
thou art thou
```

6. 绘制迭代训练的损失函数（见图 9-6），代码如下：

```
plt.plot(train_loss, 'k-')
plt.title('Sequence to Sequence Loss')
plt.xlabel('Generation')
plt.ylabel('Loss')
plt.show()
```

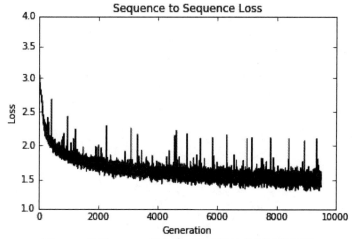

图 9-6 多层 LSTM 莎士比亚模型的迭代训练损失图

9.4.3 工作原理

TensorFlow 通过 MultiRNNCell() 函数输入 RNN 单元列表即可将 RNN 层扩展为多层 RNN。本节使用的莎士比亚文本数据集和上一节相同，但是处理时用字符而不是单词。我们将文本数据传入三层 LSTM 模型生成莎士比亚文本。发现在迭代训练 10 个 epoch 后，训练模型已经可以生成古英语的单词。

9.5 用 TensorFlow 实现 Seq2Seq 翻译模型

因为每个 RNN 单元都有输出，所以我们能训练 RNN 序列预测变长的序列。本节将利用该特性创建英语到德语的翻译模型。

9.5.1 开始

本节将构建英语到德语的翻译模型。TensorFlow 自带相关模型类来进行 Seq2Seq 翻译模型训练，我们将说明如何在所下载的英语 - 德语语句对上训练和使用该模型。我们将使用的数据来自 http://www.manythings.org/ 的 zip 文件，该文件汇编了 Tatoeba 库的数据。这些语句数据是 tab 键分割的英语 - 德语语句对（例如：hello./t hallo），并且该语句数据包含了数千种不同长度的句子。

本节代码已升级，使用了 TensorFlow 官方提供的神经机器翻译模型（https://github.com/tensorflow/nmt）。

该项目将向你展示如何下载数据，使用、修改和添加超参数，以及配置自己的数据以使用项目文件。

虽然官方教程展示了如何通过命令行进行训练，但本教程将展示如何使用提供的内部代码从头开始训练自己的模型。

9.5.2 动手做

1. 导入必要的编程库，代码如下：

```
import os
import re
import sys
import json
import math
import time
import string
import requests
import io
import numpy as np
import collections
import random
import pickle
import string
import matplotlib.pyplot as plt
import tensorflow as tf
from zipfile import ZipFile
from collections import Counter
from tensorflow.python.ops import lookup_ops
from tensorflow.python.framework import ops
ops.reset_default_graph()

local_repository = 'temp/seq2seq'
```

2. 以下代码块将整个 NMT 模型库导入 temp 文件夹：

```
if not os.path.exists(local_repository):
    from git import Repo
    tf_model_repository = 'https://github.com/tensorflow/nmt/'
    Repo.clone_from(tf_model_repository, local_repository)
    sys.path.insert(0, 'temp/seq2seq/nmt/')

# May also try to use 'attention model' by importing the attention model:
# from temp.seq2seq.nmt import attention_model as attention_model
from temp.seq2seq.nmt import model as model
from temp.seq2seq.nmt.utils import vocab_utils as vocab_utils
import temp.seq2seq.nmt.model_helper as model_helper
import temp.seq2seq.nmt.utils.iterator_utils as iterator_utils
import temp.seq2seq.nmt.utils.misc_utils as utils
import temp.seq2seq.nmt.train as train
```

3. 接下来设置关于词汇量大小的参数、删除哪些标点符号以及数据存储位置：

```
# Model Parameters
vocab_size = 10000
punct = string.punctuation

# Data Parameters
```

```
data_dir = 'temp'
data_file = 'eng_ger.txt'
model_path = 'seq2seq_model'
full_model_dir = os.path.join(data_dir, model_path)
```

4. 我们将使用 TensorFlow 提供的超参数格式。这种类型的参数存储（json 或 xml 格式文件）允许我们在不同类型架构（在不同的文件中）中重复使用。对于本例，我们将使用提供给我们的 wmt16.json 并进行一些变化：

```
# Load hyper-parameters for translation model. (Good defaults are 
provided in Repository).
hparams = tf.contrib.training.HParams()
param_file = 'temp/seq2seq/nmt/standard_hparams/wmt16.json'
# Can also try: (For different architectures)
# 'temp/seq2seq/nmt/standard_hparams/iwslt15.json'
# 'temp/seq2seq/nmt/standard_hparams/wmt16_gnmt_4_layer.json',
# 'temp/seq2seq/nmt/standard_hparams/wmt16_gnmt_8_layer.json',

with open(param_file, "r") as f:
    params_json = json.loads(f.read())

for key, value in params_json.items():
    hparams.add_hparam(key, value)
hparams.add_hparam('num_gpus', 0)
hparams.add_hparam('num_encoder_layers', hparams.num_layers)
hparams.add_hparam('num_decoder_layers', hparams.num_layers)
hparams.add_hparam('num_encoder_residual_layers', 0)
hparams.add_hparam('num_decoder_residual_layers', 0)
hparams.add_hparam('init_op', 'uniform')
hparams.add_hparam('random_seed', None)
hparams.add_hparam('num_embeddings_partitions', 0)
hparams.add_hparam('warmup_steps', 0)
hparams.add_hparam('length_penalty_weight', 0)
hparams.add_hparam('sampling_temperature', 0.0)
hparams.add_hparam('num_translations_per_input', 1)
hparams.add_hparam('warmup_scheme', 't2t')
hparams.add_hparam('epoch_step', 0)
hparams.num_train_steps = 5000

# Not use any pretrained embeddings
hparams.add_hparam('src_embed_file', '')
hparams.add_hparam('tgt_embed_file', '')
hparams.add_hparam('num_keep_ckpts', 5)
hparams.add_hparam('avg_ckpts', False)

# Remove attention
hparams.attention = None
```

5. 如果模型和数据目录不存在，则创建它们：

```
# Make Model Directory
if not os.path.exists(full_model_dir):
    os.makedirs(full_model_dir)

# Make data directory
if not os.path.exists(data_dir):
    os.makedirs(data_dir)
```

6. 下载数据集，并将数据拆分为英语和德语句子的单词列表：

```
print('Loading English-German Data')
# Check for data, if it doesn't exist, download it and save it
if not os.path.isfile(os.path.join(data_dir, data_file)):
    print('Data not found, downloading Eng-Ger sentences from www.manythings.org')
    sentence_url = 'http://www.manythings.org/anki/deu-eng.zip'
    r = requests.get(sentence_url)
    z = ZipFile(io.BytesIO(r.content))
    file = z.read('deu.txt')
    # Format Data
    eng_ger_data = file.decode('utf-8')
    eng_ger_data = eng_ger_data.encode('ascii', errors='ignore')
    eng_ger_data = eng_ger_data.decode().split('\n')
    # Write to file
    with open(os.path.join(data_dir, data_file), 'w') as out_conn:
        for sentence in eng_ger_data:
            out_conn.write(sentence + '\n')
else:
    eng_ger_data = []
    with open(os.path.join(data_dir, data_file), 'r') as in_conn:
        for row in in_conn:
            eng_ger_data.append(row[:-1])
print('Done!')
```

7. 现在我们删除英语和德语句子的标点符号：

```
# Remove punctuation
eng_ger_data = [''.join(char for char in sent if char not in punct) for sent in eng_ger_data]
# Split each sentence by tabs
eng_ger_data = [x.split('\t') for x in eng_ger_data if len(x) >= 1]
[english_sentence, german_sentence] = [list(x) for x in zip(*eng_ger_data)]
english_sentence = [x.lower().split() for x in english_sentence]
german_sentence = [x.lower().split() for x in german_sentence]
```

8. 为了使用 TensorFlow 中更快的数据管道功能，我们需要以适当的格式将格式化的数据写入磁盘。翻译模型期望的格式如下：

```
train_prefix.source_suffix = train.en
 train_prefix.target_suffix = train.de
```

后缀确定语言（en = English，de = deutsch），前缀确定数据集的类型（训练集或测试集）：

```
# We need to write them to separate text files for the text-line-dataset operations.
train_prefix = 'train'
src_suffix = 'en' # English
tgt_suffix = 'de' # Deutsch (German)
source_txt_file = train_prefix + '.' + src_suffix
hparams.add_hparam('src_file', source_txt_file)
target_txt_file = train_prefix + '.' + tgt_suffix
hparams.add_hparam('tgt_file', target_txt_file)
with open(source_txt_file, 'w') as f:
    for sent in english_sentence:
        f.write(' '.join(sent) + '\n')

with open(target_txt_file, 'w') as f:
```

```
        for sent in german_sentence:
            f.write(' '.join(sent) + '\n')
```

9. 解析一些测试句子。我们任意选择大约 100 个句子，然后也将它们写入文件：

```
# Partition some sentences off for testing files
test_prefix = 'test_sent'
hparams.add_hparam('dev_prefix', test_prefix)
hparams.add_hparam('train_prefix', train_prefix)
hparams.add_hparam('test_prefix', test_prefix)
hparams.add_hparam('src', src_suffix)
hparams.add_hparam('tgt', tgt_suffix)

num_sample = 100
total_samples = len(english_sentence)
# Get around 'num_sample's every so often in the src/tgt sentences
ix_sample = [x for x in range(total_samples) if x % (total_samples
// num_sample) == 0]
test_src = [' '.join(english_sentence[x]) for x in ix_sample]
test_tgt = [' '.join(german_sentence[x]) for x in ix_sample]

# Write test sentences to file
with open(test_prefix + '.' + src_suffix, 'w') as f:
    for eng_test in test_src:
        f.write(eng_test + '\n')

with open(test_prefix + '.' + tgt_suffix, 'w') as f:
    for ger_test in test_src:
        f.write(ger_test + '\n')
```

10. 处理英语和德语句子的词汇表，然后将词汇列表保存到文件中：

```
print('Processing the vocabularies.')
# Process the English Vocabulary
all_english_words = [word for sentence in english_sentence for word
in sentence]
all_english_counts = Counter(all_english_words)
eng_word_keys = [x[0] for x in
all_english_counts.most_common(vocab_size-3)] # -3 because UNK, S,
/S is also in there
eng_vocab2ix = dict(zip(eng_word_keys, range(1, vocab_size)))
eng_ix2vocab = {val: key for key, val in eng_vocab2ix.items()}
english_processed = []
for sent in english_sentence:
    temp_sentence = []
    for word in sent:
        try:
            temp_sentence.append(eng_vocab2ix[word])
        except KeyError:
            temp_sentence.append(0)
    english_processed.append(temp_sentence)

# Process the German Vocabulary
all_german_words = [word for sentence in german_sentence for word
in sentence]
all_german_counts = Counter(all_german_words)
ger_word_keys = [x[0] for x in
all_german_counts.most_common(vocab_size-3)]
```

```python
# -3 because UNK, S, /S is also in there
ger_vocab2ix = dict(zip(ger_word_keys, range(1, vocab_size)))
ger_ix2vocab = {val: key for key, val in ger_vocab2ix.items()}
german_processed = []
for sent in german_sentence:
    temp_sentence = []
    for word in sent:
        try:
            temp_sentence.append(ger_vocab2ix[word])
        except KeyError:
            temp_sentence.append(0)
    german_processed.append(temp_sentence)


# Save vocab files for data processing
source_vocab_file = 'vocab' + '.' + src_suffix
hparams.add_hparam('src_vocab_file', source_vocab_file)
eng_word_keys = ['<unk>', '<s>', '</s>'] + eng_word_keys

target_vocab_file = 'vocab' + '.' + tgt_suffix
hparams.add_hparam('tgt_vocab_file', target_vocab_file)
ger_word_keys = ['<unk>', '<s>', '</s>'] + ger_word_keys

# Write out all unique english words
with open(source_vocab_file, 'w') as f:
    for eng_word in eng_word_keys:
        f.write(eng_word + '\n')

# Write out all unique german words
with open(target_vocab_file, 'w') as f:
    for ger_word in ger_word_keys:
        f.write(ger_word + '\n')

# Add vocab size to hyper parameters
hparams.add_hparam('src_vocab_size', vocab_size)
hparams.add_hparam('tgt_vocab_size', vocab_size)

# Add out-directory
out_dir = 'temp/seq2seq/nmt_out'
hparams.add_hparam('out_dir', out_dir)
if not tf.gfile.Exists(out_dir):
    tf.gfile.MakeDirs(out_dir)
```

11. 分别创建训练、推理和评价计算图。首先创建训练计算图，我们用类实现，并将参数集成到namedtuple()中。此代码来自TensorFlow的NMT库，更多信息，请参考该库中model_helper.py文件：

```
class TrainGraph(collections.namedtuple("TrainGraph", ("graph",
"model", "iterator", "skip_count_placeholder"))):
    pass

def create_train_graph(scope=None):
    graph = tf.Graph()
    with graph.as_default():
        src_vocab_table, tgt_vocab_table =
vocab_utils.create_vocab_tables(hparams.src_vocab_file,
hparams.tgt_vocab_file,share_vocab=False)
```

```python
        src_dataset = tf.data.TextLineDataset(hparams.src_file)
        tgt_dataset = tf.data.TextLineDataset(hparams.tgt_file)
        skip_count_placeholder = tf.placeholder(shape=(),
dtype=tf.int64)

        iterator = iterator_utils.get_iterator(src_dataset,
tgt_dataset, src_vocab_table, tgt_vocab_table,
batch_size=hparams.batch_size, sos=hparams.sos, eos=hparams.eos,
random_seed=None, num_buckets=hparams.num_buckets,
src_max_len=hparams.src_max_len, tgt_max_len=hparams.tgt_max_len,
skip_count=skip_count_placeholder)

        final_model = model.Model(hparams, iterator=iterator,
mode=tf.contrib.learn.ModeKeys.TRAIN,
source_vocab_table=src_vocab_table,
target_vocab_table=tgt_vocab_table, scope=scope)

        return TrainGraph(graph=graph, model=final_model,
iterator=iterator, skip_count_placeholder=skip_count_placeholder)

train_graph = create_train_graph()
```

12. 创建评价图：

```python
# Create the evaluation graph
class EvalGraph(collections.namedtuple("EvalGraph", ("graph",
"model", "src_file_placeholder",
"tgt_file_placeholder","iterator"))):
    pass

def create_eval_graph(scope=None):
    graph = tf.Graph()

    with graph.as_default():
        src_vocab_table, tgt_vocab_table =
vocab_utils.create_vocab_tables(
            hparams.src_vocab_file, hparams.tgt_vocab_file,
hparams.share_vocab)
        src_file_placeholder = tf.placeholder(shape=(),
dtype=tf.string)
        tgt_file_placeholder = tf.placeholder(shape=(),
dtype=tf.string)
        src_dataset = tf.data.TextLineDataset(src_file_placeholder)
        tgt_dataset = tf.data.TextLineDataset(tgt_file_placeholder)
        iterator = iterator_utils.get_iterator(
            src_dataset,
            tgt_dataset,
            src_vocab_table,
            tgt_vocab_table,
            hparams.batch_size,
            sos=hparams.sos,
            eos=hparams.eos,
            random_seed=hparams.random_seed,
            num_buckets=hparams.num_buckets,
            src_max_len=hparams.src_max_len_infer,
            tgt_max_len=hparams.tgt_max_len_infer)
        final_model = model.Model(hparams,
                                   iterator=iterator,
```

```
                mode=tf.contrib.learn.ModeKeys.EVAL,
                source_vocab_table=src_vocab_table,
                target_vocab_table=tgt_vocab_table,
                                    scope=scope)
        return EvalGraph(graph=graph,
                        model=final_model,
                        src_file_placeholder=src_file_placeholder,
                        tgt_file_placeholder=tgt_file_placeholder,
                        iterator=iterator)

eval_graph = create_eval_graph()
```

13. 创建推理图：

```
# Inference graph
class InferGraph(collections.namedtuple("InferGraph",
("graph","model","src_placeholder",
"batch_size_placeholder","iterator"))):
    pass

def create_infer_graph(scope=None):
    graph = tf.Graph()
    with graph.as_default():
        src_vocab_table, tgt_vocab_table =
vocab_utils.create_vocab_tables(hparams.src_vocab_file,hparams.tgt_
vocab_file, hparams.share_vocab)
        reverse_tgt_vocab_table =
lookup_ops.index_to_string_table_from_file(hparams.tgt_vocab_file,
default_value=vocab_utils.UNK)

        src_placeholder = tf.placeholder(shape=[None],
dtype=tf.string)
        batch_size_placeholder = tf.placeholder(shape=[],
dtype=tf.int64)
        src_dataset =
tf.data.Dataset.from_tensor_slices(src_placeholder)
        iterator = iterator_utils.get_infer_iterator(src_dataset,
src_vocab_table,
batch_size=batch_size_placeholder,
eos=hparams.eos,
src_max_len=hparams.src_max_len_infer)
        final_model = model.Model(hparams,
                                    iterator=iterator,
mode=tf.contrib.learn.ModeKeys.INFER,
source_vocab_table=src_vocab_table,
target_vocab_table=tgt_vocab_table,
reverse_target_vocab_table=reverse_tgt_vocab_table,
                                    scope=scope)
        return InferGraph(graph=graph,
                        model=final_model,
                        src_placeholder=src_placeholder,
batch_size_placeholder=batch_size_placeholder,
                        iterator=iterator)

infer_graph = create_infer_graph()
```

14. 为了在训练期间提供更有解释性的输出，我们提供一些在训练期间能作为输出的源

语言 / 目标语言对，代码如下：

```
# Create sample data for evaluation
sample_ix = [25, 125, 240, 450]
sample_src_data = [' '.join(english_sentence[x]) for x in
sample_ix]
sample_tgt_data = [' '.join(german_sentence[x]) for x in sample_ix]
print([x for x in zip(sample_src_data, sample_tgt_data)])
```

15. 加载训练图：

```
config_proto = utils.get_config_proto()

train_sess = tf.Session(config=config_proto,
graph=train_graph.graph)
eval_sess = tf.Session(config=config_proto, graph=eval_graph.graph)
infer_sess = tf.Session(config=config_proto,
graph=infer_graph.graph)

# Load the training graph
with train_graph.graph.as_default():
    loaded_train_model, global_step =
model_helper.create_or_load_model(train_graph.model,
hparams.out_dir,
train_sess,
"train")

summary_writer =
tf.summary.FileWriter(os.path.join(hparams.out_dir, 'Training'),
train_graph.graph)
```

16. 现在将评价操作添加到计算图中：

```
for metric in hparams.metrics:
    hparams.add_hparam("best_" + metric, 0)
    best_metric_dir = os.path.join(hparams.out_dir, "best_" +
metric)
    hparams.add_hparam("best_" + metric + "_dir", best_metric_dir)
    tf.gfile.MakeDirs(best_metric_dir)

eval_output = train.run_full_eval(hparams.out_dir, infer_graph,
infer_sess, eval_graph, eval_sess, hparams, summary_writer,
sample_src_data, sample_tgt_data)

eval_results, _, acc_blue_scores = eval_output
```

17. 初始化计算图，初始化每次迭代都会更新的参数（时间、全局步长和 epoch 数）：

```
# Training Initialization
last_stats_step = global_step
last_eval_step = global_step
last_external_eval_step = global_step

steps_per_eval = 10 * hparams.steps_per_stats
steps_per_external_eval = 5 * steps_per_eval
avg_step_time = 0.0
step_time, checkpoint_loss, checkpoint_predict_count = 0.0, 0.0,
0.0
checkpoint_total_count = 0.0
speed, train_ppl = 0.0, 0.0
```

```
utils.print_out("# Start step %d, lr %g, %s" %
                (global_step,
loaded_train_model.learning_rate.eval(session=train_sess),
                time.ctime()))
skip_count = hparams.batch_size * hparams.epoch_step
utils.print_out("# Init train iterator, skipping %d elements" %
skip_count)

train_sess.run(train_graph.iterator.initializer,
           feed_dict={train_graph.skip_count_placeholder:
skip_count})
```

> 请注意,默认每 1000 次迭代训练将保存一次模型。如果需要,也可以改变这个超参数。目前,训练此模型并保存最新的五个模型占用大约 2 GB 的硬盘空间。

18. 以下代码将进行模型的训练和评价。训练最重要的部分是在循环的最开始(前三分之一),其余代码专门用于评价、样本推理和保存模型。代码如下:

```
# Run training
while global_step < hparams.num_train_steps:
    start_time = time.time()
    try:
        step_result = loaded_train_model.train(train_sess)
        (_, step_loss, step_predict_count, step_summary,
global_step, step_word_count,
         batch_size, __, ___) = step_result
        hparams.epoch_step += 1
    except tf.errors.OutOfRangeError:
        # Next Epoch
        hparams.epoch_step = 0
        utils.print_out("# Finished an epoch, step %d. Perform
external evaluation" % global_step)
        train.run_sample_decode(infer_graph,
                                infer_sess,
                                hparams.out_dir,
                                hparams,
                                summary_writer,
                                sample_src_data,
                                sample_tgt_data)
        dev_scores, test_scores, _ =
train.run_external_eval(infer_graph,
infer_sess,
hparams.out_dir,
hparams,
summary_writer)
        train_sess.run(train_graph.iterator.initializer,
feed_dict={train_graph.skip_count_placeholder: 0})
        continue

    summary_writer.add_summary(step_summary, global_step)

    # Statistics
    step_time += (time.time() - start_time)
    checkpoint_loss += (step_loss * batch_size)
```

```
        checkpoint_predict_count += step_predict_count
        checkpoint_total_count += float(step_word_count)

        # print statistics
        if global_step - last_stats_step >= hparams.steps_per_stats:
            last_stats_step = global_step
            avg_step_time = step_time / hparams.steps_per_stats
            train_ppl = utils.safe_exp(checkpoint_loss / checkpoint_predict_count)
            speed = checkpoint_total_count / (1000 * step_time)

            utils.print_out(" global step %d lr %g "
                            "step-time %.2fs wps %.2fK ppl %.2f %s" %
                            (global_step,
loaded_train_model.learning_rate.eval(session=train_sess),
                             avg_step_time, speed, train_ppl,
train._get_best_results(hparams))))

            if math.isnan(train_ppl):
                break

            # Reset timer and loss.
            step_time, checkpoint_loss, checkpoint_predict_count = 0.0, 0.0, 0.0
            checkpoint_total_count = 0.0

        if global_step - last_eval_step >= steps_per_eval:
            last_eval_step = global_step
            utils.print_out("# Save eval, global step %d" % global_step)
            utils.add_summary(summary_writer, global_step, "train_ppl", train_ppl)

            # Save checkpoint
            loaded_train_model.saver.save(train_sess,
os.path.join(hparams.out_dir, "translate.ckpt"),
global_step=global_step)

            # Evaluate on dev/test
            train.run_sample_decode(infer_graph,
                                    infer_sess,
                                    out_dir,
                                    hparams,
                                    summary_writer,
                                    sample_src_data,
                                    sample_tgt_data)
            dev_ppl, test_ppl = train.run_internal_eval(eval_graph,
                                                        eval_sess,
                                                        out_dir,
                                                        hparams,
                                                        summary_writer)

        if global_step - last_external_eval_step >= steps_per_external_eval:
            last_external_eval_step = global_step

            # Save checkpoint
            loaded_train_model.saver.save(train_sess,
```

```
              os.path.join(hparams.out_dir, "translate.ckpt"),
              global_step=global_step)
            train.run_sample_decode(infer_graph,
                                    infer_sess,
                                    out_dir,
                                    hparams,
                                    summary_writer,
                                    sample_src_data,
                                    sample_tgt_data)
            dev_scores, test_scores, _ =
    train.run_external_eval(infer_graph,
    infer_sess,
    out_dir,
    hparams,
    summary_writer)
```

9.5.3　工作原理

本节使用 TensorFlow 内建的 Seq2Seq 模型将英语翻译为德语。

虽然我们的测试句子并没有得到很好的翻译效果，但是仍有提升的空间。如果模型训练时间加长，合并一些桶（每个桶中放更多训练数据），就可以提高翻译水平。

9.5.4　延伸学习

在 manythings 网站上也有其他语言的双语语料库（http://www.manythings.org/anki/），读者可以下载使用其他语料库。

9.6　TensorFlow 训练孪生 RNN 度量相似度

相对于其他算法模型来讲，RNN 模型的一大优点就是能处理变长的序列数据。利用该特性以及其可以生成全新的序列数据，我们能创建方法度量输入序列和其他序列之间的相似度。本节将训练孪生 RNN 模型度量记录地址之间的相似度。

9.6.1　开始

在本节中，我们将构建一个双向 RNN 模型，加入一个全联接层，全联接层输出固定长度的数值型向量。我们创建一个双向 RNN 层，输入地址，并将输出传入全联接层，该全联接层输出固定长度的数值型向量（长度为 100）。然后比较两个向量输出的余弦相似度，其值在 –1 和 1 之间。输入数据与目标相似则为 1，否则为 –1。余弦距离的预测仅仅为输出结果的符号值（负值意味着不相似，正值意味着相似）。我们能利用该网络进行记录匹配，取基准地址与查询地址之间余弦距离最高的。图 9-7 是网络架构图。

该模型的优点是，可以接受从未见过的输入并进行比较，输出 –1 到 1。我们将在代码中选择从未见过的地址进行测试，查看模型是否能匹配相似的地址。

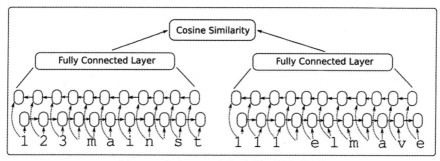

图 9-7 孪生 RNN 相似度模型的架构图

9.6.2 动手做

1. 导入必要的编程库，创建一个计算图会话，代码如下：

```
import os
import random
import string
import numpy as np
import matplotlib.pyplot as plt
import tensorflow as tf
sess = tf.Session()
```

2. 设置模型参数，代码如下：

```
batch_size = 200
n_batches = 300
max_address_len = 20
margin = 0.25
num_features = 50
dropout_keep_prob = 0.8
```

3. 创建一个孪生 RNN 相似度模型类，代码如下：

```
def snn(address1, address2, dropout_keep_prob,
        vocab_size, num_features, input_length):
    # Define the Siamese double RNN with a fully connected layer at the end
    def Siamese_nn(input_vector, num_hidden):
        cell_unit = tf.nn.rnn_cell.BasicLSTMCell
        # Forward direction cell
        lstm_forward_cell = cell_unit(num_hidden, forget_bias=1.0)
        lstm_forward_cell = tf.nn.rnn_cell.DropoutWrapper(lstm_forward_cell, output_keep_prob=dropout_keep_prob)
        # Backward direction cell
        lstm_backward_cell = cell_unit(num_hidden, forget_bias=1.0)
        lstm_backward_cell = tf.nn.rnn_cell.DropoutWrapper(lstm_backward_cell, output_keep_prob=dropout_keep_prob)
        # Split title into a character sequence
        input_embed_split = tf.split(1, input_length, input_vector)
        input_embed_split = [tf.squeeze(x, squeeze_dims=[1]) for x in input_embed_split]
```

```
        # Create bidirectional layer
        outputs, _, _ = tf.nn.bidirectional_rnn(lstm_forward_cell,
                                                lstm_backward_cell,
                                                input_embed_split,
                                                dtype=tf.float32)
        # Average The output over the sequence
        temporal_mean = tf.add_n(outputs) / input_length
        # Fully connected layer
        output_size = 10
        A = tf.get_variable(name="A", shape=[2*num_hidden, output_size],
                            dtype=tf.float32,
initializer=tf.random_normal_initializer(stddev=0.1))
        b = tf.get_variable(name="b", shape=[output_size], dtype=tf.float32,
initializer=tf.random_normal_initializer(stddev=0.1))
        final_output = tf.matmul(temporal_mean, A) + b
        final_output = tf.nn.dropout(final_output, dropout_keep_prob)
        return(final_output)
    with tf.variable_scope("Siamese") as scope:
            output1 = Siamese_nn(address1, num_features)
            # Declare that we will use the same variables on the second string
            scope.reuse_variables()
            output2 = Siamese_nn(address2, num_features)
# Unit normalize the outputs
output1 = tf.nn.l2_normalize(output1, 1)
output2 = tf.nn.l2_normalize(output2, 1)
# Return cosine distance
#   in this case, the dot product of the norms is the same.
dot_prod = tf.reduce_sum(tf.mul(output1, output2), 1)
return dot_prod
```

使用tf.variable_scope可在Siamese网络的两个部分共享变量参数。注意，余弦距离是归一化向量的点积。

4. 声明预测函数，该函数是余弦距离的符号值，代码如下：

```
def get_predictions(scores):
    predictions = tf.sign(scores, name="predictions")
    return predictions
```

5. 声明损失函数。我们希望为error预留一个margin（类似于SVM模型）。损失函数项中包括正损失和负损失。具体代码如下：

```
def loss(scores, y_target, margin):
    # Calculate the positive losses
    pos_loss_term = 0.25 * tf.square(tf.sub(1., scores))
    pos_mult = tf.cast(y_target, tf.float32)
    # Make sure positive losses are on similar strings
    positive_loss = tf.mul(pos_mult, pos_loss_term)
    # Calculate negative losses, then make sure on dissimilar strings
    neg_mult = tf.sub(1., tf.cast(y_target, tf.float32))
```

```
        negative_loss = neg_mult*tf.square(scores)
        # Combine similar and dissimilar losses
        loss = tf.add(positive_loss, negative_loss)
        # Create the margin term.  This is when the targets are 0, and
the scores are less than m, return 0.
        # Check if target is zero (dissimilar strings)
        target_zero = tf.equal(tf.cast(y_target, tf.float32), 0.)
        # Check if cosine outputs is smaller than margin
        less_than_margin = tf.less(scores, margin)
        # Check if both are true
        both_logical = tf.logical_and(target_zero, less_than_margin)
        both_logical = tf.cast(both_logical, tf.float32)
        # If both are true, then multiply by (1-1)=0.
        multiplicative_factor = tf.cast(1. - both_logical, tf.float32)
        total_loss = tf.mul(loss, multiplicative_factor)
        # Average loss over batch
        avg_loss = tf.reduce_mean(total_loss)
        return avg_loss
```

6. 声明准确度函数，代码如下：

```
def accuracy(scores, y_target):
    predictions = get_predictions(scores)
    correct_predictions = tf.equal(predictions, y_target)
    accuracy = tf.reduce_mean(tf.cast(correct_predictions,
tf.float32))
    return accuracy
```

7. 使用基准地址创建有"打印错误"的相似地址，我们将标注基准地址和错误地址，代码如下：

```
def create_typo(s):
    rand_ind = random.choice(range(len(s)))
    s_list = list(s)
    s_list[rand_ind]=random.choice(string.ascii_lowercase +
'0123456789')
    s = ''.join(s_list)
    return s
```

8. 将街道号、街道名和街道后缀随机组合生成数据。街道名和街道后缀的列表如下：

```
street_names = ['abbey', 'baker', 'canal', 'donner', 'elm',
'fifth', 'grandvia', 'hollywood', 'interstate', 'jay', 'kings']
street_types = ['rd', 'st', 'ln', 'pass', 'ave', 'hwy', 'cir',
'dr', 'jct']
```

9. 生成测试查询地址和基准地址，代码如下：

```
test_queries = ['111 abbey ln', '271 doner cicle',
                '314 king avenue', 'tensorflow is fun']
test_references = ['123 abbey ln', '217 donner cir', '314 kings
ave', '404 hollywood st', 'tensorflow is so fun']
```

最后的查询和基准地址对于本例模型来说都未见过，但是我们希望模型能识别出它们的相似性。

10. 定义如何生成批量数据。本例的批量数据是一半相似的地址（基准地址和"打印错误"地址）和一半不相似的地址。不相似的地址是通过读取地址列表的后半部分，并使用 numpy.roll() 函数将其向右循环移动 1 位获取的，代码如下：

```
def get_batch(n):
    # Generate a list of reference addresses with similar addresses that have
    # a typo.
    numbers = [random.randint(1, 9999) for i in range(n)]
    streets = [random.choice(street_names) for i in range(n)]
    street_suffs = [random.choice(street_types) for i in range(n)]
    full_streets = [str(w) + ' ' + x + ' ' + y for w,x,y in zip(numbers, streets, street_suffs)]
    typo_streets = [create_typo(x) for x in full_streets]
    reference = [list(x) for x in zip(full_streets, typo_streets)]
    # Shuffle last half of them for training on dissimilar addresses
    half_ix = int(n/2)
    bottom_half = reference[half_ix:]
    true_address = [x[0] for x in bottom_half]
    typo_address = [x[1] for x in bottom_half]
    typo_address = list(np.roll(typo_address, 1))
    bottom_half = [[x,y] for x,y in zip(true_address, typo_address)]
    reference[half_ix:] = bottom_half
    # Get target similarities (1's for similar, -1's for non-similar)
    target = [1]*(n-half_ix) + [-1]*half_ix
    reference = [[x,y] for x,y in zip(reference, target)]
    return reference
```

11. 定义地址词汇表，以及如何将地址 one-hot 编码为索引，代码如下：

```
vocab_chars = string.ascii_lowercase + '0123456789 '
vocab2ix_dict = {char:(ix+1) for ix, char in enumerate(vocab_chars)}
vocab_length = len(vocab_chars) + 1

# Define vocab one-hot encoding
def address2onehot(address,
                   vocab2ix_dict = vocab2ix_dict,
                   max_address_len = max_address_len):
    # translate address string into indices
    address_ix = [vocab2ix_dict[x] for x in list(address)]
    # Pad or crop to max_address_len
    address_ix = (address_ix + [0]*max_address_len)[0:max_address_len]
    return address_ix
```

12. 处理好词汇表，我们开始声明模型占位符和嵌入查找函数。对于嵌入查找来说，我们将使用 one-hot 编码嵌入，使用单位矩阵作为查找矩阵，代码如下：

```
address1_ph = tf.placeholder(tf.int32, [None, max_address_len], name="address1_ph")
address2_ph = tf.placeholder(tf.int32, [None, max_address_len], name="address2_ph")
y_target_ph = tf.placeholder(tf.int32, [None], name="y_target_ph")
```

```
dropout_keep_prob_ph = tf.placeholder(tf.float32,
name="dropout_keep_prob")

# Create embedding lookup
identity_mat = tf.diag(tf.ones(shape=[vocab_length]))
address1_embed = tf.nn.embedding_lookup(identity_mat, address1_ph)
address2_embed = tf.nn.embedding_lookup(identity_mat, address2_ph)
```

13. 声明算法模型、准确度、损失函数和预测操作，代码如下：

```
# Define Model
text_snn = model.snn(address1_embed, address2_embed,
dropout_keep_prob_ph,
                    vocab_length, num_features, max_address_len)
# Define Accuracy
batch_accuracy = model.accuracy(text_snn, y_target_ph)
# Define Loss
batch_loss = model.loss(text_snn, y_target_ph, margin)
# Define Predictions
predictions = model.get_predictions(text_snn)
```

14. 在开始训练模型之前，增加优化器函数，初始化变量，代码如下：

```
# Declare optimizer
optimizer = tf.train.AdamOptimizer(0.01)
# Apply gradients
train_op = optimizer.minimize(batch_loss)
# Initialize Variables
init = tf.global_variables_initializer()
sess.run(init)
```

15. 现在遍历迭代训练，记录损失函数和准确度，代码如下：

```
train_loss_vec = []
train_acc_vec = []
for b in range(n_batches):
    # Get a batch of data
    batch_data = get_batch(batch_size)
    # Shuffle data
    np.random.shuffle(batch_data)
    # Parse addresses and targets
    input_addresses = [x[0] for x in batch_data]
    target_similarity = np.array([x[1] for x in batch_data])
    address1 = np.array([address2onehot(x[0]) for x in
input_addresses])
    address2 = np.array([address2onehot(x[1]) for x in
input_addresses])
    train_feed_dict = {address1_ph: address1,
                       address2_ph: address2,
                       y_target_ph: target_similarity,
                       dropout_keep_prob_ph: dropout_keep_prob}

    _, train_loss, train_acc = sess.run([train_op, batch_loss,
batch_accuracy],
                                        feed_dict=train_feed_dict)
    # Save train loss and accuracy
    train_loss_vec.append(train_loss)
    train_acc_vec.append(train_acc)
```

16. 训练模型之后，我们处理测试查询和基准地址来查看模型效果，代码如下：

```
    test_queries_ix = np.array([address2onehot(x) for x in
test_queries])
    test_references_ix = np.array([address2onehot(x) for x in
test_references])
    num_refs = test_references_ix.shape[0]
    best_fit_refs = []
    for query in test_queries_ix:
        test_query = np.repeat(np.array([query]), num_refs, axis=0)
        test_feed_dict = {address1_ph: test_query,
                          address2_ph: test_references_ix,
                          y_target_ph: target_similarity,
                          dropout_keep_prob_ph: 1.0}
        test_out = sess.run(text_snn, feed_dict=test_feed_dict)
        best_fit = test_references[np.argmax(test_out)]
        best_fit_refs.append(best_fit)
    print('Query Addresses: {}'.format(test_queries))
    print('Model Found Matches: {}'.format(best_fit_refs))
```

17. 输出结果如下：

```
Query Addresses: ['111 abbey ln', '271 doner cicle', '314 king
avenue', 'tensorflow is fun']
Model Found Matches: ['123 abbey ln', '217 donner cir', '314 kings
ave', 'tensorflow is so fun']
```

9.6.3 延伸学习

从测试查询和基准地址可以看出，训练的模型不仅能判别正确的基准地址，也可以生成非地址词语。通过训练模型的损失和准确度来看下模型的效果（见图9-9）。

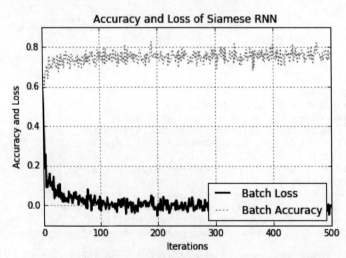

图 9-8 孪生 RNN 相似度模型训练的准确度和损失趋势图

注意，本例没有指定测试集来练习，因为本例是生成数据。我们创建批量函数每次创建新的批量数据，所以模型输入的都是新数据，用批量损失和准确度代替测试损失和准确度。但是对于实际的有限数据集来讲，并不一定都正确，所以我们需要训练集和测试集来评估模型效果。

第 10 章

TensorFlow 产品化

本章将介绍以下知识点：
- TensorFlow 的单元测试
- TensorFlow 的多设备使用
- 分布式 TensorFlow 实践
- TensorFlow 产品化开发提示
- TensorFlow 产品化的实例
- TensorFlow 服务部署

10.1 简介

到目前为止，我们已学习了如何使用 TensorFlow 训练和评估各种模型。本章将展示如何编写产品化使用的代码。产品级代码的定义有很多，这里我们将产品级代码定义为有单元测试的代码、分开训练和评估代码、有效地存储和加载数据管道中各种所需的部分以及创建计算图会话。

 本章提供的 Python 脚本需在命令行运行。测试运行后会在屏幕上打印日志。

10.2 TensorFlow 的单元测试

测试代码使得原型设计更快、调试更有效、调整更快、代码分享也更容易。本节 TensorFlow 的单元测试实现相当容易。

10.2.1 开始

当编写一个 TensorFlow 模型时，单元测试帮助我们测试代码功能。当我们想调整代码单元时，测试将确保这些改变不会破坏模型。本节将在 MNIST 数据集上创建一个简单的

CNN 网络,并实现三种不同的单元测试,以此介绍如何在 TensorFlow 中编写单元测试。

 注意,Python 有一个名为 Nose 的测试库。TensorFlow 也有内建的测试函数,我们将展示如何使用这些函数测试张量对象的值,并且没有在计算图会话中评估该值。

1. 导入必要的编程库,并格式化数据集,代码如下:

```
import sys
import numpy as np
import tensorflow as tf
from tensorflow.python.framework import ops
ops.reset_default_graph()
# Start a graph session
sess = tf.Session()
# Load data
data_dir = 'temp'
mnist = tf.keras.datasets.mnist
(train_xdata, train_labels), (test_xdata, test_labels) = mnist.load_data()
train_xdata = train_xdata / 255.0
test_xdata = test_xdata / 255.0
# Set model parameters
batch_size = 100
learning_rate = 0.005
evaluation_size = 100
image_width = train_xdata[0].shape[0]
image_height = train_xdata[0].shape[1]
target_size = max(train_labels) + 1
num_channels = 1 # greyscale = 1 channel
generations = 100
eval_every = 5
conv1_features = 25
conv2_features = 50
max_pool_size1 = 2 # NxN window for 1st max pool layer
max_pool_size2 = 2 # NxN window for 2nd max pool layer
fully_connected_size1 = 100
dropout_prob = 0.75
```

2. 声明模型的占位符、变量和模型表达式,代码如下:

```
# Declare model placeholders
x_input_shape = (batch_size, image_width, image_height, num_channels)
x_input = tf.placeholder(tf.float32, shape=x_input_shape)
y_target = tf.placeholder(tf.int32, shape=(batch_size))
eval_input_shape = (evaluation_size, image_width, image_height, num_channels)
eval_input = tf.placeholder(tf.float32, shape=eval_input_shape)
eval_target = tf.placeholder(tf.int32, shape=(evaluation_size))
dropout = tf.placeholder(tf.float32, shape=())
# Declare model parameters
conv1_weight = tf.Variable(tf.truncated_normal([4, 4, num_channels, conv1_features],
                                                stddev=0.1,
dtype=tf.float32))
```

```python
conv1_bias = tf.Variable(tf.zeros([conv1_features],
dtype=tf.float32))
conv2_weight = tf.Variable(tf.truncated_normal([4, 4,
conv1_features, conv2_features],
                                                stddev=0.1,
dtype=tf.float32))
conv2_bias = tf.Variable(tf.zeros([conv2_features],
dtype=tf.float32))
# fully connected variables
resulting_width = image_width // (max_pool_size1 * max_pool_size2)
resulting_height = image_height // (max_pool_size1 *
max_pool_size2)
full1_input_size = resulting_width * resulting_height *
conv2_features
full1_weight = tf.Variable(tf.truncated_normal([full1_input_size,
fully_connected_size1],
                          stddev=0.1, dtype=tf.float32))
full1_bias =
tf.Variable(tf.truncated_normal([fully_connected_size1],
stddev=0.1, dtype=tf.float32))
full2_weight =
tf.Variable(tf.truncated_normal([fully_connected_size1,
target_size],
                                                 stddev=0.1,
dtype=tf.float32))
full2_bias = tf.Variable(tf.truncated_normal([target_size],
stddev=0.1, dtype=tf.float32))

# Initialize Model Operations
def my_conv_net(input_data):
    # First Conv-ReLU-MaxPool Layer
    conv1 = tf.nn.conv2d(input_data, conv1_weight, strides=[1, 1,
1, 1], padding='SAME')
    relu1 = tf.nn.relu(tf.nn.bias_add(conv1, conv1_bias))
    max_pool1 = tf.nn.max_pool(relu1, ksize=[1, max_pool_size1,
max_pool_size1, 1],
                               strides=[1, max_pool_size1,
max_pool_size1, 1], padding='SAME')
    # Second Conv-ReLU-MaxPool Layer
    conv2 = tf.nn.conv2d(max_pool1, conv2_weight, strides=[1, 1, 1,
1], padding='SAME')
    relu2 = tf.nn.relu(tf.nn.bias_add(conv2, conv2_bias))
    max_pool2 = tf.nn.max_pool(relu2, ksize=[1, max_pool_size2,
max_pool_size2, 1],
                               strides=[1, max_pool_size2,
max_pool_size2, 1], padding='SAME')
    # Transform Output into a 1xN layer for next fully connected
layer
    final_conv_shape = max_pool2.get_shape().as_list()
    final_shape = final_conv_shape[1] * final_conv_shape[2] *
final_conv_shape[3]
    flat_output = tf.reshape(max_pool2, [final_conv_shape[0],
final_shape])
    # First Fully Connected Layer
    fully_connected1 = tf.nn.relu(tf.add(tf.matmul(flat_output,
full1_weight), full1_bias))
    # Second Fully Connected Layer
```

```
        final_model_output = tf.add(tf.matmul(fully_connected1,
full2_weight), full2_bias)
        # Add dropout
        final_model_output = tf.nn.dropout(final_model_output, dropout)
        return final_model_output

model_output = my_conv_net(x_input)
test_model_output = my_conv_net(eval_input)
```

3. 创建损失函数以及预测操作和准确度操作，然后初始化模型变量，代码如下：

```
# Declare Loss Function (softmax cross entropy)
loss =
tf.reduce_mean(tf.nn.sparse_softmax_cross_entropy_with_logits(model
_output, y_target))
# Create a prediction function
prediction = tf.nn.softmax(model_output)
test_prediction = tf.nn.softmax(test_model_output)

# Create accuracy function
def get_accuracy(logits, targets):
    batch_predictions = np.argmax(logits, axis=1)
    num_correct = np.sum(np.equal(batch_predictions, targets))
    return 100. * num_correct/batch_predictions.shape[0]

# Create an optimizer
my_optimizer = tf.train.MomentumOptimizer(learning_rate, 0.9)
train_step = my_optimizer.minimize(loss)
# Initialize Variables
init = tf.global_variables_initializer()
sess.run(init)
```

4. 使用 tf.test.TestCase() 类来测试占位符或者变量的值。在本例中，我们要确保 dropout 概率大于 0.25，因此模型训练时并不会试图让 dropout 大于 75%，代码如下：

```
# Check values of tensors!
class DropOutTest(tf.test.TestCase):
    # Make sure that we don't drop too much
    def dropout_greaterthan(self):
        with self.test_session():
            self.assertGreater(dropout.eval(), 0.25)
```

5. 测试准确度函数的功能正确。我们按预期创建一个样本数组，确保测试返回 100% 的准确度，代码如下：

```
# Test accuracy function
class AccuracyTest(tf.test.TestCase):
    # Make sure accuracy function behaves correctly
    def accuracy_exact_test(self):
        with self.test_session():
            test_preds = [[0.9, 0.1],[0.01, 0.99]]
            test_targets = [0, 1]
test_acc = get_accuracy(test_preds, test_targets)
self.assertEqual(test_acc.eval(), 100.)
```

6. 测试张量的形状符合预期。测试模型输出结果是预期的 [batch_size, target_size] 形状，代码如下：

```
# Test tensorshape
class ShapeTest(tf.test.TestCase):
    # Make sure our model output is size [batch_size, num_classes]
    def output_shape_test(self):
        with self.test_session():
            numpy_array = np.ones([batch_size, target_size])
            self.assertShapeEqual(numpy_array, model_output)
```

7. 创建 main() 函数来测试正在运行的 TensorFlow，代码如下：

```
def main(argv):
    # Start training loop
    train_loss = []
    train_acc = []
    test_acc = []
    for i in range(generations):
        rand_index = np.random.choice(len(train_xdata), size=batch_size)
        rand_x = train_xdata[rand_index]
        rand_x = np.expand_dims(rand_x, 3)
        rand_y = train_labels[rand_index]
        train_dict = {x_input: rand_x, y_target: rand_y, dropout: dropout_prob}

        sess.run(train_step, feed_dict=train_dict)
        temp_train_loss, temp_train_preds = sess.run([loss, prediction], feed_dict=train_dict)
        temp_train_acc = get_accuracy(temp_train_preds, rand_y)

        if (i + 1) % eval_every == 0:
            eval_index = np.random.choice(len(test_xdata), size=evaluation_size)
            eval_x = test_xdata[eval_index]
            eval_x = np.expand_dims(eval_x, 3)
            eval_y = test_labels[eval_index]
            test_dict = {eval_input: eval_x, eval_target: eval_y, dropout: 1.0}
            test_preds = sess.run(test_prediction, feed_dict=test_dict)
            temp_test_acc = get_accuracy(test_preds, eval_y)
            # Record and print results
            train_loss.append(temp_train_loss)
            train_acc.append(temp_train_acc)
            test_acc.append(temp_test_acc)
            acc_and_loss = [(i + 1), temp_train_loss, temp_train_acc, temp_test_acc]
            acc_and_loss = [np.round(x, 2) for x in acc_and_loss]
            print('Generation # {}. Train Loss: {:.2f}. Train Acc (Test Acc): {:.2f} ({:.2f})'.format(*acc_and_loss))
```

8. 为了让我们的程序实现训练或测试，需要在命令行调用它，以下代码是主程序代码，如果输入参数 test，则执行测试，否则执行训练：

```
if __name__ == '__main__':
    cmd_args = sys.argv
    if len(cmd_args) > 1 and cmd_args[1] == 'test':
```

```
        # Perform unit-tests
        tf.test.main(argv=cmd_args[1:])
    else:
        # Run the TensorFlow app
        tf.app.run(main=None, argv=cmd_args)
```

9. 如果在命令行运行此程序，应该会得到如下输出：

```
$ python3 implementing_unit_tests.py test
...
----------------------------------------------------------------------
Ran 3 tests in 0.001s

OK
```

完整的代码可以在本书的 GitHub 地址（https://github.com/nfmcclure/tensorflow_cookbook/ 或 https://github.com/PacktPublishing/TensorFlow-Machine-Learning-Cookbook- Second-Edition）获取。

10.2.2 工作原理

本节实现了三种单元测试：张量值、操作输出结果和张量形状。关于 TensorFlow 更多的单元测试函数请浏览 https://www.tensorflow.org/versions/master/api_docs/python/test.html。

注意，单元测试可以帮助我们确保代码的功能符合预期，提供分享代码的信心，并使代码可重用。

10.3 TensorFlow 的多设备使用

TensorFlow 有很多特点，其中计算图非常适合并行计算。就像处理不同批量一样，计算图可以由不同的处理器来处理。本节将介绍如何在同一台机器上访问不同的处理器。

10.3.1 开始

本节将展示如何在同一系统上访问多个设备，并训练算法模型。有一个很常见的现象，单 CPU 经常配备一个或多个可分享计算能力的 GPU。如果 TensorFlow 能访问这些设备，通过贪婪策略它将自动在多个设备上分布计算。但是 TensorFlow 也允许程序通过命名空间的方式指定哪个操作将在哪个设备上执行。

为了访问 GPU 设备，需要安装 TensorFlow 的 GPU 版。安装 TensorFlow 的 GPU 版请访问：https://www. tensorflow. org/ versions/ master/get_ started/ os_ setup. html，并根据自身操作系统选择安装步骤。值得一提的是，TensorFlow 的 GPU 版要求安装 CUDA 来使用 GPU。

本节将介绍一系列的命令来访问操作系统上的各种设备，并找出哪个设备可供 TensorFlow 使用。

10.3.2 动手做

1. 为了能够确定 TensorFlow 的什么操作正在使用什么设备，我们在计算图会话中传入一个 config 参数，将 log_device_placement 设为 True。当我们在命令行运行脚本时，会看到指定设备布局，代码如下：

```
import tensorflow as tf
sess = tf.Session(config=tf.ConfigProto(log_device_placement=True))
a = tf.constant([1.0, 2.0, 3.0, 4.0, 5.0, 6.0], shape=[2, 3],
name='a')
b = tf.constant([1.0, 2.0, 3.0, 4.0, 5.0, 6.0], shape=[3, 2],
name='b')
c = tf.matmul(a, b)
# Runs the op.
print(sess.run(c))
```

2. 从控制台运行下面的命令：

```
$python3 using_multiple_devices.py
Device mapping: no known devices.
I tensorflow/core/common_runtime/direct_session.cc:175] Device
mapping:
MatMul: /job:localhost/replica:0/task:0/cpu:0
I tensorflow/core/common_runtime/simple_placer.cc:818] MatMul:
/job:localhost/replica:0/task:0/cpu:0
b: /job:localhost/replica:0/task:0/cpu:0
I tensorflow/core/common_runtime/simple_placer.cc:818] b:
/job:localhost/replica:0/task:0/cpu:0
a: /job:localhost/replica:0/task:0/cpu:0
I tensorflow/core/common_runtime/simple_placer.cc:818] a:
/job:localhost/replica:0/task:0/cpu:0
[[ 22.  28.]
 [ 49.  64.]]
```

3. TensorFlow 会自动在计算设备（CPU 和 GPU）上分配计算，但是有时我们希望搞清楚 TensorFlow 正在使用的设备。当加载先前保存过的模型，并且该模型在计算图中已分配固定设备时，本地机器可提供不同的设备给计算图使用。实现该功能只需在 config 设置设备，代码如下：

```
config = tf.ConfigProto()
config.allow_soft_placement = True
sess_soft = tf.Session(config=config)
```

4. 当使用 GPU 时，TensorFlow 默认占据大部分 GPU 内存。虽然这也是时常期望的，但是我们也要谨慎分配 GPU 内存。当 TensorFlow 一直不释放 GPU 内存时，如有必要，我们可以设置 GPU 内存增长选项让 GPU 内存分配缓慢增大到最大限制，代码如下：

```
config.gpu_options.allow_growth = True
sess_grow = tf.Session(config=config)
```

5. 如果希望限制死 TensorFlow 使用 GPU 内存的百分比，可以使用 config 设置 per_process_gpu_memory_fraction，代码如下：

```
config.gpu_options.per_process_gpu_memory_fraction = 0.4
sess_limited = tf.Session(config=config)
```

6. 有时，我们希望代码健壮到可以确定哪些 GPU 可用。TensorFlow 的内建函数可以实现该功能。如果我们期望代码在 GPU 内存合适时利用 GPU 计算能力，并分配指定操作给 GPU，那么该功能是有益的。具体代码如下：

```
if tf.test.is_built_with_cuda():
    <Run GPU specific code here>
```

7. 我们希望分配指定操作给 GPU。下面是一个示例代码，做了一些简单的计算，并将它们分配给主 CPU 和两个副 GPU，代码如下：

```
with tf.device('/cpu:0'):
    a = tf.constant([1.0, 3.0, 5.0], shape=[1, 3])
    b = tf.constant([2.0, 4.0, 6.0], shape=[3, 1])
    with tf.device('/gpu:0'):
        c = tf.matmul(a,b)
        c = tf.reshape(c, [-1])
    with tf.device('/gpu:1'):
        d = tf.matmul(b,a)
        flat_d = tf.reshape(d, [-1])
    combined = tf.multiply(c, flat_d)
print(sess.run(combined))
```

10.3.3 工作原理

当希望为 TensorFlow 的操作指定机器的特定设备，我们需要知道如何引用该设备。在 TensorFlow 中，设备约定俗成的命名如下表所示。

设　备	设备名
主 CPU	/cpu:0
副 CPU	/cpu:1
主 GPU	/gpu:0
副 GPU	/gpu:1
第三 GPU	/gpu:2

10.3.4 延伸学习

TensorFlow 在云服务上运行越来越容易。许多云服务提供商提供带有主 CPU 和 GPU 的 GPU 实例。亚马逊（Amazon Web Services，AWS）的 G 实例和 P2 实例提供 GPU 计算能力加速 TensorFlow 的处理。还有 AWS 机器学习镜像（AWS Machine Image，AMI），其可免费启动已安装 TensorFlow 的 GPU 实例。

10.4 分布式 TensorFlow 实践

为了扩展 TensorFlow 的并行能力，可将分开的计算图操作以分布式方式运行在不同的机器上。本节主要介绍分布式 TensorFlow 实践。

10.4.1 开始

TensorFlow 发布之后几个月，Google 公司发布了分布式 TensorFlow。这是 TensorFlow 生态系统的一个巨大提升，允许构建一个 TensorFlow 集群（分布于不同的 worker 机器），共享相同的训练和评估模型的计算任务。使用分布式的 TensorFlow 只需要简单地设置 worker 节点参数，然后为不同的 worker 节点分配不同的作业。

本节将建立两个本地 worker 并分配不同的作业。

10.4.2 动手做

1. 加载 TensorFlow，定义两个本地 worker（端口分别为 2222 和 2223），代码如下：

```
import tensorflow as tf
# Cluster for 2 local workers (tasks 0 and 1):
cluster = tf.train.ClusterSpec({'local': ['localhost:2222',
'localhost:2223']})
```

2. 将两个 worker 加入到集群中，并标记 task 数字，代码如下：

```
server = tf.train.Server(cluster, job_name="local", task_index=0)
server = tf.train.Server(cluster, job_name="local", task_index=1)
```

3. 现在我们为每个 worker 分配一个 task。第一个 worker 将初始化两个矩阵（每个是 25×25）。第二个 worker 计算每个矩阵所有元素的和。然后自动分配将两个和求和的任务，并打印出结果，代码如下：

```
mat_dim = 25
matrix_list = {}
with tf.device('/job:local/task:0'):
    for i in range(0, 2):
        m_label = 'm_{}'.format(i)
        matrix_list[m_label] = tf.random_normal([mat_dim, mat_dim])
# Have each worker calculate the sums
sum_outs = {}
with tf.device('/job:local/task:1'):
    for i in range(0, 2):
        A = matrix_list['m_{}'.format(i)]
        sum_outs['m_{}'.format(i)] = tf.reduce_sum(A)
    # Sum all the sums
    summed_out = tf.add_n(list(sum_outs.values()))
with tf.Session(server.target) as sess:
    result = sess.run(summed_out)
    print('Summed Values:{}'.format(result))
```

4. 运行下面的命令：

```
$ python3 parallelizing_tensorflow.py
I tensorflow/core/distributed_runtime/rpc/grpc_channel.cc:197]
Initialize GrpcChannelCache for job local -> {0 -> localhost:2222,
1 -> localhost:2223}
I tensorflow/core/distributed_runtime/rpc/grpc_server_lib.cc:206]
Started server with target: grpc://localhost:2222
I tensorflow/core/distributed_runtime/rpc/grpc_channel.cc:197]
Initialize GrpcChannelCache for job local -> {0 -> localhost:2222,
```

```
1 -> localhost:2223}
I tensorflow/core/distributed_runtime/rpc/grpc_server_lib.cc:206]
Started server with target: grpc://localhost:2223
I tensorflow/core/distributed_runtime/master_session.cc:928] Start
master session 252bb6f530553002 with config:
Summed Values:-21.12611198425293
```

10.4.3 工作原理

使用分布式 TensorFlow 相当容易。我们只需在集群服务器中为 worker 节点分配带名字的 IP。然后就可以手动或者自动为 worker 节点分配操作任务。

10.5 TensorFlow 产品化开发提示

如果我们想在产品中使用机器学习的脚本，有一些最佳实践的注意点。本节将指出一些有用的最佳实践。

10.5.1 开始

本节总结提炼了 TensorFlow 产品化的注意点，包括如何最有效地保存和加载词汇表、计算图、模型变量和检查点。我们也会讨论如何使用 TensorFlow 的命令行参数解析器和日志级别。

10.5.2 动手做

1. 当运行 TensorFlow 程序时，我们可能希望确保内存中没有其他计算图会话，或者每次调试程序时重置计算图会话，代码如下：

```
from tensorflow.python.framework import ops
ops.reset_default_graph()
```

2. 当处理文本或者任意数据管道，我们需要确保保存处理过的数据，以便随后用相同的方式处理评估数据。如果处理文本数据，我们需要确定能够保存和加载词汇字典。下面的代码是保存 JSON 格式的词汇字典的例子：

```
import json word_list = ['to', 'be', 'or', 'not', 'to', 'be']
vocab_list = list(set(word_list))
vocab2ix_dict = dict(zip(vocab_list, range(len(vocab_list))))
ix2vocab_dict = {val:key for key,val in vocab2ix_dict.items()}

# Save vocabulary
import json
with open('vocab2ix_dict.json', 'w') as file_conn:
    json.dump(vocab2ix_dict, file_conn)

# Load vocabulary
with open('vocab2ix_dict.json', 'r') as file_conn:
    vocab2ix_dict = json.load(file_conn)
```

本例使用 JSON 格式存储词汇字典,但是我们也可以将其存储为 txt、CSV 或者二进制文件。如果词汇字典太大,则建议使用二进制文件。可以用 pickle 库创建 pkl 二进制文件,但是 pkl 文件在不同的 Python 版本间兼容性不好。

3. 为了保存算法模型的计算图和变量,我们在计算图中创建 Saver() 操作。建议在模型训练过程中按一定规则保存模型。具体代码如下:

```
After model declaration, add a saving operations
saver = tf.train.Saver()
# Then during training, save every so often, referencing the
training generation
for i in range(generations):
    ...
    if i%save_every == 0:
        saver.save(sess, 'my_model', global_step=step)
# Can also save only specific variables:
saver = tf.train.Saver({"my_var": my_variable})
```

注意,Saver() 操作也可以传参数。它能接收变量和张量的字典来保存指定元素,也可以接收 checkpoint_every_n_hours 参数来按时间规则保存操作。默认保存操作只保存最近的五个模型(考虑存储空间),但是也可以通过 max_to_keep 选项改变。

4. 在保存算法模型之前,确保为模型重要的操作命名。如果不提前命名,TensorFlow 很难加载指定占位符、操作或者变量。TensorFlow 的大部分操作和函数都接受 name 参数,代码如下:

```
conv_weights = tf.Variable(tf.random_normal(), name='conv_weights')
loss = tf.reduce_mean(... , name='loss')
```

5. TensorFlow 的 tf.apps.flags 库使得执行命令行参数解析相当容易。你可以定义 string、float、integer 或者 boolean 型的命令行参数。运行 tf.app.run() 函数即可运行带有 flag 参数的 main() 函数,代码如下:

```
tf.flags.DEFINE_string("worker_locations", "", "List of worker
addresses.")
tf.flags.DEFINE_float('learning_rate', 0.01, 'Initial learning
rate.')
tf.flags.DEFINE_integer('generations', 1000, 'Number of training
generations.')
tf.flags.DEFINE_boolean('run_unit_tests', False, 'If true, run
tests.')
FLAGS = tf.flags.FLAGS
# Need to define a 'main' function for the app to run
def main(_):
    worker_ips = FLAGS.worker_locations.split(",")
    learning_rate = FLAGS.learning_rate
    generations = FLAGS.generations
    run_unit_tests = FLAGS.run_unit_tests

# Run the Tensorflow app
```

```
if __name__ == "__main__":
    # The following is looking for a "main()" function to run and will pass.
    tf.app.run()
    # Can modify this to be more custom:
    tf.app.run(main=my_main_function(), argv=my_arguments)
```

6. TensorFlow 有内建的 logging 设置日志级别。其日志级别可设置为 DEBUG、INFO、WARN、ERROR 和 FATAL，默认级别是 WARN，代码如下：

```
tf.logging.set_verbosity(tf.logging.WARN)
# WARN is the default value, but to see more information, you can set it to
#    INFO or DEBUG
tf.logging.set_verbosity(tf.logging.DEBUG)
```

10.5.3 工作原理

本节提供了在 TensorFlow 中编写产品级代码的注意事项。我们希望通过介绍 app-flag 的概念、模型存储和 logging，使得读者能习惯地用这些工具编写代码，并能理解这些工具在他人代码中的运用。使用其他方法也可以写出产品级代码，下一节将展示一个完整的示例。

10.6 TensorFlow 产品化的实例

产品级机器学习模型的最佳实践是分开训练和评估代码。本节将展示评估脚本，包括单元测试、模型保存和加载，以及模型评估。

10.6.1 开始

本节将展示如何使用上面的产品级标准实现评估脚本。代码实际由训练脚本和评估脚本组成，但是本节将只讲解评估脚本。注意，两个脚本都可以在 GitHub（https://github.com/nfmcclure/tensorflow_cookbook/ 或 https://github.com/PacktPublishing/TensorFlow-Machine-Learning-Cookbook-Second-Edition）在线获取。

本例将实现第 9 章的第一个 RNN 模型例子，预测邮件文本信息是否为垃圾信息。我们假设 RNN 模型已训练好并保存，同时带有词汇表。

10.6.2 动手做

1. 导入必要的编程库，声明 TensorFlow 应用的 flag，代码如下：

```
import os
import re
import numpy as np
import tensorflow as tf
from tensorflow.python.framework import ops
ops.reset_default_graph()
# Define App Flags
```

```
tf.flags.DEFINE_string("storage_folder", "temp", "Where to store
model and data.")
tf.flags.DEFINE_float('learning_rate', 0.0005, 'Initial learning
rate.')
tf.flags.DEFINE_float('dropout_prob', 0.5, 'Per to keep probability
for dropout.')
tf.flags.DEFINE_integer('epochs', 20, 'Number of epochs for
training.')
tf.flags.DEFINE_integer('batch_size', 250, 'Batch Size for
training.')
tf.flags.DEFINE_integer('rnn_size', 15, 'RNN feature size.')
tf.flags.DEFINE_integer('embedding_size', 25, 'Word embedding
size.')
tf.flags.DEFINE_integer('min_word_frequency', 20, 'Word frequency
cutoff.')
tf.flags.DEFINE_boolean('run_unit_tests', False, 'If true, run
tests.')

FLAGS = tf.flags.FLAGS
```

2. 声明文本清洗函数，在训练脚本中也有相同的清洗函数，代码如下：

```
def clean_text(text_string):
    text_string = re.sub(r'([^sw]|_|[0-9])+', '', text_string)
    text_string = " ".join(text_string.split())
    text_string = text_string.lower()
    return text_string
```

3. 加载词汇处理函数，代码如下：

```
def load_vocab():
    vocab_path = os.path.join(FLAGS.storage_folder, "vocab")
    vocab_processor =
tf.contrib.learn.preprocessing.VocabularyProcessor.restore(vocab_pa
th)
    return vocab_processor
```

4. 有了清洗的文本数据和词汇处理函数，即可创建数据处理管道，代码如下：

```
def process_data(input_data, vocab_processor):
    input_data = clean_text(input_data)
    input_data = input_data.split()
    processed_input =
np.array(list(vocab_processor.transform(input_data)))
    return processed_input
```

5. 我们需要数据来评估模型。为此，我们将要求用户在屏幕上输入文本，然后处理输入文本和返回处理文本，代码如下：

```
def get_input_data():
    input_text = input("Please enter a text message to evaluate: ")
    vocab_processor = load_vocab()
    return process_data(input_text, vocab_processor)
```

对于本例而言，我们通过要求用户输入来创建评估数据，也有许多应用通过提供文件或者 API 来获取数据，我们可以根据需要调整输入函数。

6. 对于单元测试，应确保文本处理函数的表现符合预期，代码如下：

```
class clean_test(tf.test.TestCase):
    # Make sure cleaning function behaves correctly
    def clean_string_test(self):
        with self.test_session():
            test_input = '--Tensorflow\'s so Great! Dont you think so? '
            test_expected = 'tensorflows so great don you think so'
            test_out = clean_text(test_input)
            self.assertEqual(test_expected, test_out)
```

7. 现在，有了算法模型和数据集，我们运行主函数。该主函数获取数据集，建立计算图，加载模型变量，传入处理过的数据，打印输出结果，代码如下：

```
def main(args):
    # Get flags
    storage_folder = FLAGS.storage_folder
    # Get user input text
    x_data = get_input_data()
    # Load model
    graph = tf.Graph()
    with graph.as_default():
        sess = tf.Session()
        with sess.as_default():
            # Load the saved meta graph and restore variables
            saver = tf.train.import_meta_graph("{}.meta".format(os.path.join(storage_folder, "model.ckpt")))
            saver.restore(sess, os.path.join(storage_folder, "model.ckpt"))
            # Get the placeholders from the graph by name
            x_data_ph = graph.get_operation_by_name("x_data_ph").outputs[0]
            dropout_keep_prob = graph.get_operation_by_name("dropout_keep_prob").outputs[0]
            probability_outputs = graph.get_operation_by_name("probability_outputs").outputs[0]
            # Make the prediction
            eval_feed_dict = {x_data_ph: x_data, dropout_keep_prob: 1.0}
            probability_prediction = sess.run(tf.reduce_mean(probability_outputs, 0), eval_feed_dict)
            # Print output (Or save to file or DB connection?)
            print('Probability of Spam: {:.4}'.format(probability_prediction[1]))
```

8. 如下代码展示了 main() 函数或单元测试如何运行。

```
if __name__ == "__main__":
    if FLAGS.run_unit_tests:
        # Perform unit tests
        tf.test.main()
    else:
        # Run evaluation
        tf.app.run()
```

10.6.3 工作原理

对于模型评估，我们能加载带 TensorFlow app-flag 的命令行参数，加载模型和词汇处理函数，然后运行处理过的数据训练模型并进行预测。

注意，应通过命令行运行本节的脚本；在创建算法模型和词汇字典前应检测训练脚本。

10.7 TensorFlow 服务部署

本节将展示在 TensorFlow 上如何建立 RNN 模型来预测垃圾邮件和正常邮件。首先介绍如何以 protobuf 格式保存模型，然后将模型加载到本地服务器，读取端口 9000 作为输入。

10.7.1 开始

鼓励读者阅读官方文档和 TensorFlow 服务网站上的简短教程（https://www.tensorflow.org/serving/serving_basic）。

本例重新使用 9.2 节中 RNN 的大部分代码。我们将更改模型代码，以 protobuf 格式将模型保存在使用 TensorFlow 服务所需的合适文件夹结构中。

请注意，本章中的所有脚本都应该从命令行 bash 提示符执行。

升级的安装说明请访问官方安装网址：https:// www.tensorflow.org/serving/setup。正常安装就像向 Linux 源添加 gpg-key 并运行以下安装命令一样简单：

```
$ sudo apt install tensorflow-model-server
```

10.7.2 动手做

1. 导入必要的编程库，设置 TensorFlow flag，代码如下：

```
import os
import re
import io
import sys
import requests
import numpy as np
import tensorflow as tf
from zipfile import ZipFile
from tensorflow.python.framework import ops

ops.reset_default_graph()

# Define App Flags
tf.flags.DEFINE_string("storage_folder", "temp", "Where to store model and data.")
tf.flags.DEFINE_float('learning_rate', 0.0005, 'Initial learning rate.')
```

```
tf.flags.DEFINE_float('dropout_prob', 0.5, 'Per to keep probability
for dropout.')
tf.flags.DEFINE_integer('epochs', 20, 'Number of epochs for
training.')
tf.flags.DEFINE_integer('batch_size', 250, 'Batch Size for
training.')
tf.flags.DEFINE_integer('rnn_size', 15, 'RNN feature size.')
tf.flags.DEFINE_integer('embedding_size', 25, 'Word embedding
size.')
tf.flags.DEFINE_integer('min_word_frequency', 20, 'Word frequency
cutoff.')
tf.flags.DEFINE_boolean('run_unit_tests', False, 'If true, run
tests.')

FLAGS = tf.flags.FLAGS
```

2. 与以前一样继续编写脚本。为简洁起见，我们只会在训练脚本中体现不同之处，也就是我们如何保存 protobuf 模型。这一步是在训练完成后插入相应代码。

 请注意此代码与教程代码非常相似，主要区别在于模型名称、版本号，以及我们正在保存的是 RNN 而不是 CNN。

```
# Save the finished model for TensorFlow Serving (pb file)
# Here, it's our storage folder / version number
out_path =
os.path.join(tf.compat.as_bytes(os.path.join(storage_folder, '1')))
print('Exporting finished model to : {}'.format(out_path))
builder = tf.saved_model.builder.SavedModelBuilder(out_path)

# Build the signature_def_map.
classification_inputs =
tf.saved_model.utils.build_tensor_info(x_data_ph)
classification_outputs_classes =
tf.saved_model.utils.build_tensor_info(rnn_model_outputs)

classification_signature =
(tf.saved_model.signature_def_utils.build_signature_def(
inputs={tf.saved_model.signature_constants.CLASSIFY_INPUTS:
                    classification_inputs},
outputs={tf.saved_model.signature_constants.CLASSIFY_OUTPUT_CLASSES
:
                    classification_outputs_classes},
method_name=tf.saved_model.signature_constants.CLASSIFY_METHOD_NAME
))

        tensor_info_x =
tf.saved_model.utils.build_tensor_info(x_data_ph)
        tensor_info_y =
tf.saved_model.utils.build_tensor_info(y_output_ph)

        prediction_signature = (
            tf.saved_model.signature_def_utils.build_signature_def(
                inputs={'texts': tensor_info_x},
                outputs={'scores': tensor_info_y},
```

```
method_name=tf.saved_model.signature_constants.PREDICT_METHOD_NAME)
        )
        legacy_init_op = tf.group(tf.tables_initializer(),
name='legacy_init_op')
        builder.add_meta_graph_and_variables(
            sess, [tf.saved_model.tag_constants.SERVING],
            signature_def_map={
                'predict_spam': prediction_signature,
tf.saved_model.signature_constants.DEFAULT_SERVING_SIGNATURE_DEF_KE
Y:
                    classification_signature,
            },
            legacy_init_op=legacy_init_op)

    builder.save()

    print('Done exporting!')
```

3. 需要特别注意的是，TensorFlow 服务需要特定的文件或文件夹结构来加载模型。该脚本将以以下格式（图 10-1）安装文件：

上面的屏幕截图显示了所需的目录结构。其中，创建了数据目录 temp，并建立模型版本号：1。在版本号目录中，保存 protobuf 模型和变量文件夹 variables。

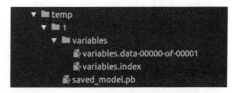

图 10-1　Tensorflow 服务期望的目录结构

 我们应该知道，在数据目录中，TensorFlow 服务将查找整数文件夹，TensorFlow 服务将自动启动并在最大整数文件夹下获取模型。这意味着要部署新模型，我们需要将其标记为版本 2，并将其粘贴到也标记为 2 的新文件夹下。然后，TensorFlow 服务将自动获取该模型。

4. 要启动服务器，需要调用命令 tensorflow_model_server，带有 port、model_name 和 model_base_path 参数。然后，TensorFlow 服务查找版本号文件夹并获取最大版本编号的模型。再通过给定参数的端口，将它部署到命令运行的机器上。在以下例子中，我们在本地机器（0.0.0.0）上运行，并且接受的默认端口为 9000：

```
$ tensorflow_model_server --port=9000 --model_name=spam_ham --
model_base_path=<directory of our
code>/tensorflow_cookbook/10_Taking_TensorFlow_to_Production/06_Usi
ng_TensorFlow_Serving/temp/

2018-08-09 12:05:16.206712: I
tensorflow_serving/model_servers/main.cc:153] Building single
TensorFlow model file config: model_name: spam_ham model_base_path:
.../temp/
2018-08-09 12:05:16.206874: I
```

```
tensorflow_serving/model_servers/server_core.cc:459]
Adding/updating models.
2018-08-09 12:05:16.206903: I
tensorflow_serving/model_servers/server_core.cc:514] (Re-)adding
model: spam_ham
2018-08-09 12:05:16.307681: I
tensorflow_serving/core/basic_manager.cc:716] Successfully reserved
resources to load servable {name: spam_ham version: 1}
2018-08-09 12:05:16.307744: I
tensorflow_serving/core/loader_harness.cc:66] Approving load for
servable version {name: spam_ham version: 1}
2018-08-09 12:05:16.307773: I
tensorflow_serving/core/loader_harness.cc:74] Loading servable
version {name: spam_ham version: 1}
2018-08-09 12:05:16.307829: I
external/org_tensorflow/tensorflow/contrib/session_bundle/bundle_sh
im.cc:360] Attempting to load native SavedModelBundle in bundle-
shim from: .../temp/1
2018-08-09 12:05:16.307867: I
external/org_tensorflow/tensorflow/cc/saved_model/loader.cc:242]
Loading SavedModel with tags: { serve }; from: .../temp/1
2018-08-09 12:05:16.313811: I
external/org_tensorflow/tensorflow/core/platform/cpu_feature_guard.
cc:141] Your CPU supports instructions that this TensorFlow binary
was not compiled to use: AVX2 FMA
2018-08-09 12:05:16.325866: I
external/org_tensorflow/tensorflow/cc/saved_model/loader.cc:161]
Restoring SavedModel bundle.
2018-08-09 12:05:16.329290: I
external/org_tensorflow/tensorflow/cc/saved_model/loader.cc:196]
Running LegacyInitOp on SavedModel bundle.
2018-08-09 12:05:16.332936: I
external/org_tensorflow/tensorflow/cc/saved_model/loader.cc:291]
SavedModel load for tags { serve }; Status: success. Took 25074
microseconds.
2018-08-09 12:05:16.332972: I
tensorflow_serving/servables/tensorflow/saved_model_warmup.cc:83]
No warmup data file found at
.../temp/1/assets.extra/tf_serving_warmup_requests
2018-08-09 12:05:16.333335: I
tensorflow_serving/core/loader_harness.cc:86] Successfully loaded
servable version {name: spam_ham version: 1}
2018-08-09 12:05:16.334678: I
tensorflow_serving/model_servers/main.cc:323] Running ModelServer
at 0.0.0.0:9000 ...
```

5. 现在可以将二进制数据提交到 <host>：9000，返回显示结果的 JSON 响应。这一步可以通过任何机器和任何编程语言来完成。不必依赖客户端拥有 TensorFlow 的本地副本是非常有用的。

10.7.3 工作原理

将之前的产品化思路与本节进行比较，主要区别在于我们在主机上部署了可以响应传入请求的模型服务器。前面的产品化思路是一个很好的使用示例，能够用于处理批处理结果或在能加载 TensorFlow 的机器上工作，但是不擅长部署可用的模型，能进行计算，并将结果返回给客户端。通过下表我们将了解两种结构的区别：

	10.6 – 批量生产	10.7 – 通过 TensorFlow 服务的生产
优点	不依赖于网络连接或主机	结果与客户端结构无关，唯一的要求是 Numpy 数组为二进制格式
缺点	客户端必须具有 TensorFlow 和模型文件	依靠主机可用
理想用途	大批量数据	生产服务始终可用，通常是小的请求

当然，每种方法的优缺点都值得商榷，两者都能满足每种情况的要求。还有许多其他可用的架构可以满足不同的需求，例如 Docker、Kubernetes、Luigi、Django / Flask、Celery、AWS 和 Azure。

10.7.4 延伸学习

本章未涉及的工具和资源的链接如下：

- **Using TensorFlow Serving in Docker:** https://www.tensorflow.org/serving/docker
- **Using TensorFlow Serving in Kubernetes:** https://www.tensorflow.org/serving/serving_inception
- **Luigi, a pipeline tool for batch jobs:** https://github.com/spotify/luigi
- **Using TensorFlow in Flask:** https://guillaumegenthial.github.io/serving.html
- **A Python framework for distributed task-queuing:** http://www.celeryproject.org/community/
- **How to use AWS lambdas with TensorFlow models:** https://aws.amazon.com/blogs/machine-learning/how-to-deploy-deep-learning-models-with-aws-lambda-and-tensorflow/

CHAPTER 11

第 11 章

TensorFlow 的进阶应用

本章将包含以下内容：
- TensorFlow 可视化：Tensorboard
- 用 TensorFlow 实现遗传算法
- 用 TensorFlow 实现 k-means 聚类算法
- 用 TensorFlow 求解常微分方程组
- 用 TensorFlow 实现随机森林算法
- 将 Keras 作为 TensorFlow API 使用

本章所有代码可以在 GitHub（https://github.com/nfmcclure/tensorflow_cookbook）或 Packt 代码库（https://github.com/PacktPublishing/TensorFlow-Machine-Learning-Cookbook-Second-Edition）获取。

11.1 简介

到目前为止，我们已经看到 TensorFlow 能够实现许多算法模型，但是 TensorFlow 能做的远不止于此。本章将展示如何使用 Tensorboard，Tensorboard 是 TensorFlow 的可视化模块，允许可视化模型训练过程中的统计指标、计算图和图像。剩余的章节将介绍如何使用 TensorFlow 的 group() 函数进行逐步更新。该函数帮助我们实现遗传算法、k-means 聚类和常微分方程（ODE）的求解，甚至创建梯度提升随机森林。

11.2 TensorFlow 可视化：Tensorboard

机器学习算法的监控和故障解决是一个令人畏惧的任务。尤其是在训练模型时，你必须等待很长时间，有时甚至不知道结果。这时使用 TensorFlow 的可视化工具 Tensorboard 可以图形化计算图，绘制模型训练中重要的值（损失、准确度和批量训练时间等）。

11.2.1 开始

为了展示 Tensorboard 的各种使用方法，本节再次实现第 3 章中的线性回归模型。我们将生成带误差的线性数据，使用 TensorFlow 的损失函数和反向传播来拟合直线。下面将介绍在 Tensorboard 中如何监控数值型数值、数据集的直方图，以及如何创建图像。

11.2.2 动手做

1. 导入必要的编程库，代码如下：

```
import os
import io
import time
import numpy as np
import matplotlib.pyplot as plt
import tensorflow as tf
```

2. 初始化计算图会话，创建 summary_writer 将 Tensorboard summary 写入 Tensorboard 文件夹，代码如下：

```
sess = tf.Session()
# Create a visualizer object
summary_writer = tf.summary.FileWriter('tensorboard', sess.graph)
```

3. 确保 summary_writer 写入的 Tensorboard 文件夹存在，以便写入 Tensorboard 日志，代码如下：

```
if not os.path.exists('tensorboard'):
    os.makedirs('tensorboard')
```

4. 设置模型参数，为模型生成线性数据集。注意，设置真实斜率 true_slope 为 2。在迭代训练时，我们将可视化斜率随时间的变化，观察它逼近真实值，代码如下：

```
batch_size = 50
generations = 100
# Create sample input data
x_data = np.arange(1000)/10.
true_slope = 2.
y_data = x_data * true_slope + np.random.normal(loc=0.0, scale=25, size=1000)
```

5. 分割数据集为测试集和训练集，代码如下：

```
train_ix = np.random.choice(len(x_data), size=int(len(x_data)*0.9), replace=False)
test_ix = np.setdiff1d(np.arange(1000), train_ix)
x_data_train, y_data_train = x_data[train_ix], y_data[train_ix]
x_data_test, y_data_test = x_data[test_ix], y_data[test_ix]
```

6. 创建占位符、变量、模型操作、损失和优化器操作，代码如下：

```
x_graph_input = tf.placeholder(tf.float32, [None])
y_graph_input = tf.placeholder(tf.float32, [None])
# Declare model variables
m = tf.Variable(tf.random_normal([1], dtype=tf.float32), name='Slope')
# Declare model
output = tf.multiply(m, x_graph_input, name='Batch_Multiplication')
# Declare loss function (L1)
```

```
residuals = output - y_graph_input
l2_loss = tf.reduce_mean(tf.abs(residuals), name="L2_Loss")
# Declare optimization function
my_optim = tf.train.GradientDescentOptimizer(0.01)
train_step = my_optim.minimize(l2_loss)
```

7. 创建 Tensorboard 操作汇总一个标量值。该汇总的标量值为模型的斜率估计，代码如下：

```
with tf.name_scope('Slope_Estimate'):
    tf.summary.scalar('Slope_Estimate', tf.squeeze(m))
```

8. 添加到 Tensorboard 的另一个汇总数据是直方图汇总。该直方图汇总输入张量，输出计算图和直方图，代码如下：

```
with tf.name_scope('Loss_and_Residuals'):
    tf.summary.histogram('Histogram_Errors', tf.squeeze(l1_loss))
    tf.summary.histogram('Histogram_Residuals',
tf.squeeze(residuals))
```

9. 创建完这些汇总操作，我们创建汇总合并操作综合所有的汇总数据，然后初始化模型变量，代码如下：

```
summary_op = tf.summary.merge_all()
# Initialize Variables
init = tf.global_variables_initializer()
sess.run(init)
```

10. 训练线性回归模型，将每次迭代训练写入汇总数据，代码如下：

```
for i in range(generations):
    batch_indices = np.random.choice(len(x_data_train), size=batch_size)
    x_batch = x_data_train[batch_indices]
    y_batch = y_data_train[batch_indices]
    _, train_loss, summary = sess.run([train_step, l2_loss, summary_op],
                                      feed_dict={x_graph_input: x_batch,
                                                 y_graph_input: y_batch})
    test_loss, test_resids = sess.run([l2_loss, residuals],
feed_dict={x_graph_input: x_data_test, y_graph_input: y_data_test})
    if (i+1)%10==0:
        print('Generation {} of {}. Train Loss: {:.3}, Test Loss: {:.3}.'.format(i+1, generations, train_loss, test_loss))

    log_writer = tf.train.SummaryWriter('tensorboard')
    log_writer.add_summary(summary, i)
```

11. 为了可视化数据点拟合的线性回归模型，我们创建 protobuf 格式的计算图图形。为此，我们创建函数输出 protobuf 格式的图形，代码如下：

```
def gen_linear_plot(slope):
    linear_prediction = x_data * slope
    plt.plot(x_data, y_data, 'b.', label='data')
    plt.plot(x_data, linear_prediction, 'r-', linewidth=3,
label='predicted line')
    plt.legend(loc='upper left')
```

```
buf = io.BytesIO()
plt.savefig(buf, format='png')
buf.seek(0)
return(buf)
```

12. 创建并且将 protobuf 格式的图形增加到 Tensorboard，代码如下：

```
# Get slope value
slope = sess.run(m)

# Generate the linear plot in buffer
plot_buf = gen_linear_plot(slope[0])

# Convert PNG buffer to TF image
image = tf.image.decode_png(plot_buf.getvalue(), channels=4)

# Add the batch dimension
image = tf.expand_dims(image, 0)

# Add image summary
image_summary_op = tf.summary.image("Linear_Plot", image)
image_summary = sess.run(image_summary_op)
log_writer.add_summary(image_summary, i)
log_writer.close()
```

注意向 Tensorboard 写入图像汇总别太频繁。例如，如果我们迭代训练 10 000 次，每次都写入汇总数据，那将生成 10 000 幅图。这会迅速占用磁盘空间。

11.2.3 延伸学习

1. 从命令行运行上面的脚本：

```
$ python3 using_tensorboard.py

Run the command: $tensorboard --logdir="tensorboard"   Then
navigate to http://127.0.0.0:6006
Generation 10 of 100. Train Loss: 20.4, Test Loss: 20.5.
Generation 20 of 100. Train Loss: 17.6, Test Loss: 20.5.
Generation 90 of 100. Train Loss: 20.1, Test Loss: 20.5.
Generation 100 of 100. Train Loss: 19.4, Test Loss: 20.5.
```

2. 运行指定的命令启动 Tensorboard，代码如下：

```
$ tensorboard --logdir="tensorboard" Starting tensorboard b'29' on
port 6006 (You can navigate to http://127.0.0.1:6006)
```

图 11-1 是 Tensorboard 看到的样例图形。可见迭代 100 次标量汇总（斜率估计）的趋势图，接近于真实值 2。

图 11-2 展示了直方图汇总的可视化，可将其视为多个折线图。

图 11-3 展示的是最终的拟合直线和数据点图形，保存为 protobuf 格式，并插入 Tensorboard 的图像汇总中。

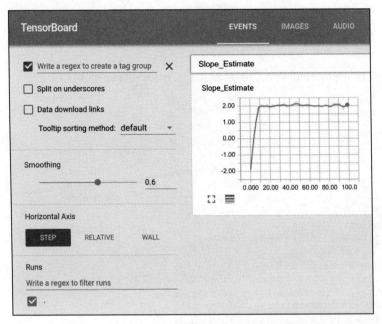

图 11-1　Tensorboard 中标量值、斜率估计的可视化

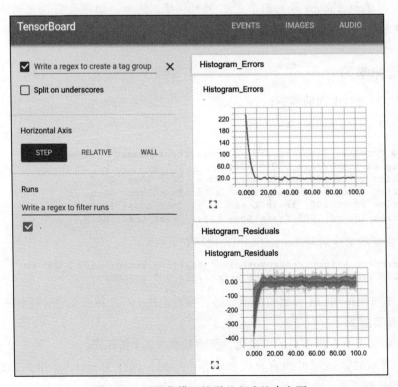

图 11-2　可视化模型的误差和残差直方图

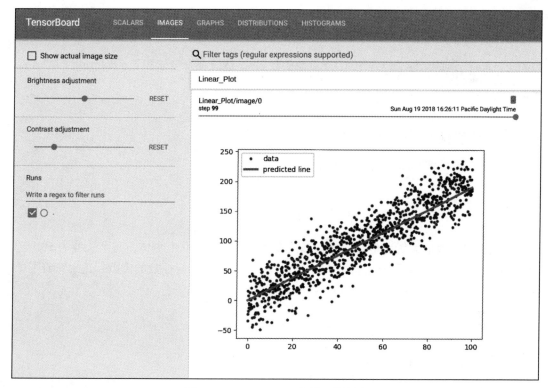

图 11-3　Tensorboard 中插入的图片

11.3　用 TensorFlow 实现遗传算法

TensorFlow 也可以用来更新任意表示为计算图的迭代型算法。迭代型算法的例子有遗传算法，用于解决最优化的问题。

11.3.1　开始

本节将介绍如何实现一个简单的遗传算法。遗传算法是最优化参数空间（离散的、连续的、光滑的、非光滑的，等等）的有效方法。基本思想是创建一个随机初始化的种群，进化、重组和突变生成新的种群（更好的质量）。通过当前种群的个体的适应度来计算各个个体的适应度。

一般来讲，遗传算法的大体步骤是：先随机初始化种群，通过各个个体的适应度排序，选择适应度较高的个体随机重组（或者交叉）创建下一代种群。下一代种群经过轻微变异产生不同于上一代的更好的适应度，然后将其放入父种群。在对子种群和父种群进行组合后，我们重复前面的整个过程。

遗传算法停止迭代的标准很多，但是本例中仅仅以迭代的总次数为标准。我们也可以在当前种群个体的适应度达到预期的标准，或者在许多次迭代后最大的适应度不再变化时，停止迭代。

本节将展示如何在 TensorFlow 中实现遗传算法。我们将生成一个最接近 ground truth 函数 $f(x) = \sin\left(\dfrac{2\pi x}{50}\right)$ 的个体（包含 50 个浮点数的数组）。适应度为个体和 ground truth 的均方误差的负值。

11.3.2 动手做

1. 导入必要的编程库，代码如下：

```
import os
import numpy as np
import matplotlib.pyplot as plt
import tensorflow as tf
```

2. 接下来，我们设置遗传算法的参数。在本例中，有 100 个个体，每个长度为 50。选择的百分比是 20%，即适应度排序前 20 的个体。变异定义为特征数的倒数，这意味着下一代种群的一个特征会变异。运行遗传算法迭代 200 次，代码如下：

```
pop_size = 100
features = 50
selection = 0.2
mutation = 1./ features
generations = 200
num_parents = int(pop_size*selection)
num_children = pop_size - num_parents
```

3. 初始化计算图会话，创建 ground truth 函数，该函数用来计算适应度，代码如下：

```
sess = tf.Session()
# Create ground truth
truth = np.sin(2*np.pi*(np.arange(features,
dtype=np.float32))/features)
```

4. 使用 TensorFlow 的变量（随机正态分布输入）初始化种群，代码如下：

```
population = tf.Variable(np.random.randn(pop_size, features),
dtype=tf.float32)
```

5. 现在创建遗传算法的占位符。占位符是为 ground truth 和每次迭代改变的数据而创建的。因为我们希望父代变化和变异概率变化交叉，这些都是模型的占位符，代码如下：

```
truth_ph = tf.placeholder(tf.float32, [1, features])
crossover_mat_ph = tf.placeholder(tf.float32, [num_children,
features])
mutation_val_ph = tf.placeholder(tf.float32, [num_children,
features])
```

6. 计算群体的适应度（均方误差的负值），选择较高适应度的个体，代码如下：

```
fitness = -tf.reduce_mean(tf.square(tf.subtract(population,
truth_ph)), 1)
top_vals, top_ind = tf.nn.top_k(fitness, k=pop_size)
```

7. 为了获得最后的结果并绘图，我们希望检索种群中适应度最高的个体，代码如下：

```
best_val = tf.reduce_min(top_vals)
best_ind = tf.argmin(top_vals, 0)
best_individual = tf.gather(population, best_ind)
```

8. 排序父种群，截取适应度较高的个体作为下一代，代码如下：

```
population_sorted = tf.gather(population, top_ind)
parents = tf.slice(population_sorted, [0, 0], [num_parents, features])
```

9. 通过创建两个随机 shuffle 的父种群矩阵来创建下一代种群。将交叉矩阵分别与 1 和 0 相加，然后与父种群矩阵相乘，生成每一代的占位符，代码如下：

```
# Indices to shuffle-gather parents
rand_parent1_ix = np.random.choice(num_parents, num_children)
rand_parent2_ix = np.random.choice(num_parents, num_children)
# Gather parents by shuffled indices, expand back out to pop_size too
rand_parent1 = tf.gather(parents, rand_parent1_ix)
rand_parent2 = tf.gather(parents, rand_parent2_ix)
rand_parent1_sel = tf.multiply(rand_parent1, crossover_mat_ph)
rand_parent2_sel = tf.multiply(rand_parent2, tf.subtract(1., crossover_mat_ph))
children_after_sel = tf.add(rand_parent1_sel, rand_parent2_sel)
```

10. 最后一个步骤是变异下一代，本例将增加一个随机正常值到下一代种群矩阵中，然后将这个矩阵和父种群连接，代码如下：

```
mutated_children = tf.add(children_after_sel, mutation_val_ph)
# Combine children and parents into new population
new_population = tf.concat(0, [parents, mutated_children])
```

11. 模型的最后一步是，使用 TensorFlow 的 group() 操作分配下一代种群到父一代种群的变量，代码如下：

```
step = tf.group(population.assign(new_population))
```

12. 初始化模型变量，代码如下：

```
init = tf.global_variables_initializer()
sess.run(init)
```

13. 迭代训练模型，再创建随机交叉矩阵和变异矩阵，更新每代的种群，代码如下：

```
for i in range(generations):
    # Create cross-over matrices for plugging in.
    crossover_mat = np.ones(shape=[num_children, features])
    crossover_point = np.random.choice(np.arange(1, features-1, step=1), num_children)
    for pop_ix in range(num_children):
        crossover_mat[pop_ix,0:crossover_point[pop_ix]]=0.
    # Generate mutation probability matrices
    mutation_prob_mat = np.random.uniform(size=[num_children, features])
    mutation_values = np.random.normal(size=[num_children, features])
    mutation_values[mutation_prob_mat >= mutation] = 0
    # Run GA step
```

```
            feed_dict = {truth_ph: truth.reshape([1, features]),
                         crossover_mat_ph: crossover_mat,
                         mutation_val_ph: mutation_values}
        step.run(feed_dict, session=sess)
        best_individual_val = sess.run(best_individual,
feed_dict=feed_dict)
        if i % 5 == 0:
            best_fit = sess.run(best_val, feed_dict = feed_dict)
            print('Generation: {}, Best Fitness (lowest MSE):
{:.2}'.format(i, -best_fit))
```

14. 输出结果如下：

```
Generation: 0, Best Fitness (lowest MSE): 1.5
Generation: 5, Best Fitness (lowest MSE): 0.83
Generation: 10, Best Fitness (lowest MSE): 0.55
Generation: 185, Best Fitness (lowest MSE): 0.085
Generation: 190, Best Fitness (lowest MSE): 0.15
Generation: 195, Best Fitness (lowest MSE): 0.083
```

11.3.3　工作原理

本节展示了如何使用TensorFlow训练简单的遗传算法。为了验证实现的遗传算法是否工作，我们绘制每次迭代适应度最高的个体和ground truth趋势图，如图11-4所示。

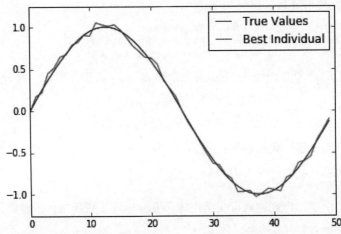

图11-4　迭代200次的适应度最高的个体和ground truth的趋势图。我们发现适应度最高的个体和ground truth相当接近

11.3.4　延伸学习

遗传算法有许多变种，可以有两个父种群和两个不同的适应度标准（例如，最低MSE和平滑度）。我们能强制变异值不大于1或者不小于–1。本例做了许多不同的调整，这些调整依赖于实际要优化的问题。对于本例来说，适应度计算相当简单，但是对于大部分遗传算法，计算适应度是一项艰巨的任务。例如，如果我们想用遗传算法优化卷积神经网络架构，

那么群体个体就有一个参数数组。参数包括过滤器大小、步长，以及卷积层。对于数据集来说，迭代固定次数后，个体的适应度是分类的准确度。如果我们在种群中有 100 个个体，那么，每次迭代必须评估 100 个不同的 CNN 模型。这是非常大强度的计算。

在使用遗传算法解决你的问题之前，需要考虑下多久才能计算完个体的适应度。如果该操作很耗时，则不适合使用遗传算法。

11.4 用 TensorFlow 实现 k-means 聚类算法

TensorFlow 也可以用来实现迭代聚类算法，比如 k-means。本节将展示在 iris 数据集中使用 k-means 算法的例子。

11.4.1 开始

本书介绍的大部分机器学习算法模型都是有监督模型，TensorFlow 是解决该类问题的理想工具。但是 TensorFlow 也可以实现无监督模型，本例将实现 k-means 聚类算法。

本例使用的数据是 iris 数据集。选择该数据集的一大好处是，我们已经知道其有三类目标（三种鸢尾花）。这让我们明确将会聚类成三个不同的类。

我们将 iris 数据集聚类成三组，然后和实际标注比较求出聚类的准确度。

11.4.2 动手做

1. 开始导入必要的编程库。因为后续需将四维的结果数据转换为二维数据进行可视化，所以也要从 sklearn 库导入 PCA 工具，代码如下：

```
import numpy as np
import matplotlib.pyplot as plt
import tensorflow as tf
from sklearn import datasets
from scipy.spatial import cKDTree
from sklearn.decomposition import PCA
from sklearn.preprocessing import scale
```

2. 创建一个计算图会话，加载 iris 数据集，代码如下：

```
sess = tf.Session()
iris = datasets.load_iris()
num_pts = len(iris.data)
num_feats = len(iris.data[0])
```

3. 设置分类数、迭代次数，创建计算图所需的变量，代码如下：

```
k=3
generations = 25
data_points = tf.Variable(iris.data)
cluster_labels = tf.Variable(tf.zeros([num_pts], dtype=tf.int64))
```

4. 声明每个分组所需的几何中心变量。我们通过随机选择 iris 数据集中的三个数据点来初始化 k-means 聚类算法的几何中心，代码如下：

```
rand_starts = np.array([iris.data[np.random.choice(len(iris.data))]
for _ in range(k)])
centroids = tf.Variable(rand_starts)
```

5. 计算每个数据点到每个几何中心的距离。本例的计算方法是，将几何中心点和数据点分别放入矩阵中，然后计算两个矩阵的欧式距离，代码如下：

```
centroid_matrix = tf.reshape(tf.tile(centroids, [num_pts, 1]),
[num_pts, k, num_feats])
point_matrix = tf.reshape(tf.tile(data_points, [1, k]), [num_pts,
k, num_feats])
distances = tf.reduce_sum(tf.square(point_matrix -
centroid_matrix), reduction_indices=2)
```

6. 分配时，是以到每个数据点最小距离为最接近的几何中心点，代码如下：

```
centroid_group = tf.argmin(distances, 1)
```

7. 计算每组分类的平均距离得到新的几何中心点，代码如下：

```
def data_group_avg(group_ids, data):
    # Sum each group
    sum_total = tf.unsorted_segment_sum(data, group_ids, 3)
    # Count each group
    num_total = tf.unsorted_segment_sum(tf.ones_like(data),
group_ids, 3)
    # Calculate average
    avg_by_group = sum_total/num_total
    return(avg_by_group)
means = data_group_avg(centroid_group, data_points)
update = tf.group(centroids.assign(means),
cluster_labels.assign(centroid_group))
```

8. 初始化模型变量，代码如下：

```
init = tf.global_variables_initializer()
sess.run(init)
```

9. 遍历迭代训练，相应地更新每组分类的几何中心点，代码如下：

```
for i in range(generations):
    print('Calculating gen {}, out of {}.'.format(i, generations))
    _, centroid_group_count = sess.run([update, centroid_group])
    group_count = []
    for ix in range(k):
        group_count.append(np.sum(centroid_group_count==ix))
    print('Group counts: {}'.format(group_count))
```

10. 输出结果如下：

```
Calculating gen 0, out of 25. Group counts: [50, 28, 72]
Calculating gen 1, out of 25. Group counts: [50, 35, 65]
Calculating gen 23, out of 25. Group counts: [50, 38, 62]
Calculating gen 24, out of 25. Group counts: [50, 38, 62]
```

11. 为了验证聚类模型，我们使用距离模型预测。看下有多少数据点与实际 iris 数据集中的鸢尾花物种匹配，代码如下：

```
[centers, assignments] = sess.run([centroids, cluster_labels])
def most_common(my_list):
    return(max(set(my_list), key=my_list.count))
label0 = most_common(list(assignments[0:50]))
label1 = most_common(list(assignments[50:100]))
label2 = most_common(list(assignments[100:150]))
group0_count = np.sum(assignments[0:50]==label0)
group1_count = np.sum(assignments[50:100]==label1)
group2_count = np.sum(assignments[100:150]==label2)
accuracy = (group0_count + group1_count + group2_count)/150.
print('Accuracy: {:.2}'.format(accuracy))
```

12. 输出结果如下：

Accuracy: 0.89

13. 为了可视化分组过程，以及是否分离出鸢尾花物种，我们将用 PCA 工具将四维结果数据转为二维结果数据，并绘制数据点和分组。PCA 分解之后，我们创建预测，并在 x-y 轴网格绘制彩色图形，代码如下：

```
pca_model = PCA(n_components=2)
reduced_data = pca_model.fit_transform(iris.data)
# Transform centers
reduced_centers = pca_model.transform(centers)
# Step size of mesh for plotting
h = .02
x_min, x_max = reduced_data[:, 0].min() - 1, reduced_data[:, 0].max() + 1
y_min, y_max = reduced_data[:, 1].min() - 1, reduced_data[:, 1].max() + 1
xx, yy = np.meshgrid(np.arange(x_min, x_max, h), np.arange(y_min, y_max, h))
# Get k-means classifications for the grid points
xx_pt = list(xx.ravel())
yy_pt = list(yy.ravel())
xy_pts = np.array([[x,y] for x,y in zip(xx_pt, yy_pt)])
mytree = cKDTree(reduced_centers)
dist, indexes = mytree.query(xy_pts)
indexes = indexes.reshape(xx.shape)
```

14. 下面是用 matplotlib 模块在同一幅图形中绘制所有结果的代码。绘图部分的代码来自 scikit-learn 官方文档的示例（http://scikit-learn.org/stable/auto_examples/cluster/plot_kmeans_digits.html），代码如下：

```
plt.clf()
plt.imshow(indexes, interpolation='nearest',
          extent=(xx.min(), xx.max(), yy.min(), yy.max()),
          cmap=plt.cm.Paired,
          aspect='auto', origin='lower')
# Plot each of the true iris data groups
symbols = ['o', '^', 'D']
label_name = ['Setosa', 'Versicolour', 'Virginica']
for i in range(3):
    temp_group = reduced_data[(i*50):(50)*(i+1)]
```

```
    plt.plot(temp_group[:, 0], temp_group[:, 1], symbols[i],
markersize=10, label=label_name[i])
# Plot the centroids as a white X
plt.scatter(reduced_centers[:, 0], reduced_centers[:, 1],
            marker='x', s=169, linewidths=3,
            color='w', zorder=10)
plt.title('K-means clustering on Iris Datasets'
          'Centroids are marked with white cross')
plt.xlim(x_min, x_max)
plt.ylim(y_min, y_max)
plt.legend(loc='lower right')
plt.show()
```

图 11-5 展示三个类、三个类的预测空间以及每类的几何中心。

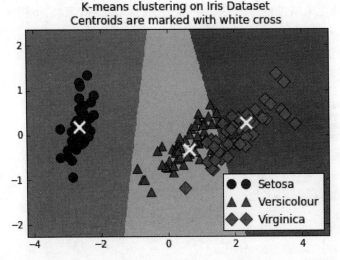

图 11-5 展示 k-means 聚类算法如何进行无监督分类，本例聚合为三类鸢尾花。三类 k-means 分组为图中三种阴影区域，三种不同的数据点（圆形、三角形和菱形）是三个鸢尾花物种

11.4.3 延伸学习

本节使用 TensorFlow 将 iris 数据集聚类为三组。然后计算数据点落入分组的百分比（89%），绘制 k-means 分组的图形。因为 k-means 算法作为分类算法的一种，其为局部线性，所以很难学习到杂色鸢尾花（versicolour）和维吉尼亚鸢尾花（virginica）之间的非线性边界。但是 k-means 算法的优点是无须标注数据集。

11.5 用 TensorFlow 求解常微分方程组

TensorFlow 可以实现许多算法，其中一个很好的例子是实现常微分方程（ODE）的求解器。求解一个常微分方程数值解是一个很容易用计算图表达的迭代过程。本节将求解洛特卡－沃尔泰拉（Lotka-Volterra）掠食者－猎物方程组。

11.5.1 开始

本节将介绍如何求解常微分方程组。我们可以使用前面两节的方法来更新迭代数值，求解常微分方程组。

本例考虑的著名常微分方程组是洛特卡－沃尔泰拉掠食者－猎物方程组。经常用该方程组描述生物系统中掠食者与猎物进行互动时的动态模型。

1920年，洛特卡－沃尔泰拉方程组发布在一篇文章中，见11.2.3节图11-1。我们使用文章中相似的参数来描述周期性系统，其离散数学表达式为：

$$X_{t+1} = X_t + (aX_t + bX_tY_t)\Delta t$$
$$X_{t+1} = X_t + (cY_t + dX_tY_t)\Delta t$$

其中，X是猎物，Y是掠食者。我们通过a、b、c和d的值来决定哪个是猎物，哪个是掠食者。对于猎物，$a>0$，$b<0$；对于掠食者，$c<0$，$d>0$。我们使用TensorFlow实现其离散版的求解器。

11.5.2 动手做

1. 导入必要的编程库，创建一个计算图会话，代码如下：

```
import matplotlib.pyplot as plt
import tensorflow as tf
sess = tf.Session()
```

2. 声明计算图中的常量和变量，代码如下：

```
x_initial = tf.constant(1.0)
y_initial = tf.constant(1.0)
X_t1 = tf.Variable(x_initial)
Y_t1 = tf.Variable(y_initial)
# Make the placeholders
t_delta = tf.placeholder(tf.float32, shape=())
a = tf.placeholder(tf.float32, shape=())
b = tf.placeholder(tf.float32, shape=())
c = tf.placeholder(tf.float32, shape=())
d = tf.placeholder(tf.float32, shape=())
```

3. 实现前面介绍的离散方程组，然后更新X和Y，代码如下：

```
X_t2 = X_t1 + (a * X_t1 + b * X_t1 * Y_t1) * t_delta
Y_t2 = Y_t1 + (c * Y_t1 + d * X_t1 * Y_t1) * t_delta
# Update to New Population
step = tf.group(
  X_t1.assign(X_t2),
  Y_t1.assign(Y_t2))
```

4. 初始化计算图，运行离散常微分方程组，用指定参数说明周期性变化，代码如下：

```
init = tf.global_variables_initializer() sess.run(init) # Run the
ODE prey_values = [] predator_values = [] for i in range(1000): #
Step simulation (using constants for a known cyclic solution)
step.run({a: (2./3.), b: (-4./3.), c: -1.0, d: 1.0, t_delta: 0.01},
session=sess) # Store each outcome temp_prey, temp_pred =
sess.run([X_t1, Y_t1]) prey_values.append(temp_prey)
predator_values.append(temp_pred)
```

> 获得洛特卡-沃尔泰拉方程的稳定求解与指定参数和起始值有较大关系。我们鼓励读者尝试不同的参数和初始值来看看会发生什么。

5. 现在绘制掠食者与猎物的值（见图11-6），代码如下：

```
plt.plot(prey_values, label="Prey")
plt.plot(predator_values, label="Predator")
plt.legend(loc='upper right')
plt.show()
```

11.5.3 工作原理

使用TensorFlow实现常微分方程组的离散版求解器。对于指定参数，我们看到掠食者与猎物的生态系统确实是周期性变化的。这在生物生态系统中是有意义的，因为如果有太多的掠食者，猎物就会濒临灭绝；缺少猎物，掠食者只能获得更少的食物，那掠食者也会开始濒临灭绝。

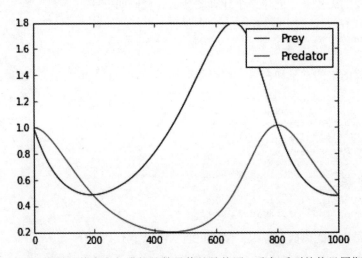

图11-6 ODE求解的掠食者与猎物的数量值的趋势图。我们看到趋势呈周期性变化

11.5.4 参考

Lotka, A. J., *Analytical note on certain rhythmic relations in organic systems.* Proc. Nat. Acad. 6 (1920) (https://www.ncbi.nlm.nih.gov/pmc/articles/PMC1084562/).

11.6 用 TensorFlow 实现随机森林算法

随机森林算法集成了多个决策树，决策树的观测值和特征都有可能是随机选择的。我们不会介绍如何训练决策树，但会给出一些能够使用梯度提升来训练的随机森林类型，可以用 TensorFlow 来计算随机森林算法。

11.6.1 开始

基于树的算法传统上是非平滑的，因为它们对数据进行分区以最小化目标输出中的方差。非平滑方法不适合基于梯度的方法。TensorFlow 依赖以下事实：模型中使用的函数是平滑的，并且它自动计算如何更改模型参数以最小化函数损失。TensorFlow 克服这个缺陷的方式是对决策边界进行平滑逼近，我们可以使用 softmax 函数或类似形状函数来逼近决策边界。

决策树将通过生成规则对数据集进行硬拆分，例如，如果 $x > 0.5$，则移动到树的这个分支，这告诉我们整个数据子集是组合在一起还是拆分，这取决于 x 的值。对决策树的平滑逼近处理的是概率而不是拆分，这意味着数据集的每次观测都有一定概率存在于树的每个末端节点中。图 11-7 对传统决策树和概率决策树进行了比较，更好地说明了这些差异。

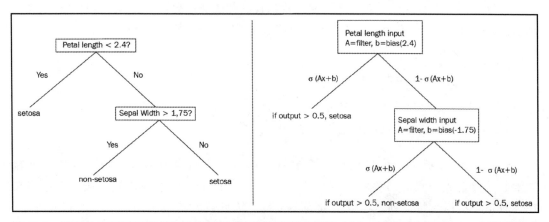

图 11-7　该图给出了不可微的标准决策树（左）和平滑决策树（右），说明了使用 sigmoid 函数能生成出现在标记叶或末端节点中观测的概率

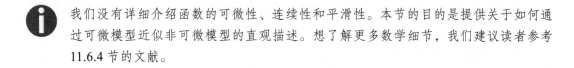

我们没有详细介绍函数的可微性、连续性和平滑性。本节的目的是提供关于如何通过可微模型近似非可微模型的直观描述。想了解更多数学细节，我们建议读者参考 11.6.4 节的文献。

11.6.2 动手做

TensorFlow 包含了本节所依赖的一些默认模型估计函数。有两个主要的梯度提升模型：

回归树和分类树。本例中我们将使用回归树来预测波士顿房屋价格数据集（https://www.cs.toronto.edu/~delve/data/boston/bostonDetail.html）。

1. 导入必要的编程库：

```
import numpy as np
import tensorflow as tf
from keras.datasets import boston_housing
from tensorflow.python.framework import ops
ops.reset_default_graph()
```

2. 从TensorFlow估计器库中设置我们将要使用的模型。本例使用BoostedTreesRegressor模型，该模型用于使用梯度提升树的回归，代码如下：

```
regression_classifier = tf.estimator.BoostedTreesRegressor
```

 对于二值分类问题，读者可以使用BoostedTreesClassifier估计器。目前不支持多类别分类，尽管将来会支持。

3. 使用Keras数据导入函数加载波士顿房屋价格数据集，如下所示：

```
(x_train, y_train), (x_test, y_test) = boston_housing.load_data()
```

4. 设置一些模型参数。批量大小是一次迭代的训练数据数量，我们将进行500次迭代训练，梯度提升森林将有100棵树，每棵树的最大深度（分裂数量）为6。

```
# Batch size
batch_size = 32
# Number of training steps
train_steps = 500
# Number of trees in our 'forest'
n_trees = 100
# Maximum depth of any tree in forest
max_depth = 6
```

5. TensorFlow提供的模型估计器需要输入函数。我们将为估计器函数创建数据输入函数。但首先，我们需要将数据放入正确标记的numpy数组格式的字典中。这些在TensorFlow中称为特征列。纯数字列尚不支持，因此我们将数字列放入自动存储桶中，如下所示：(a) 二值特征将具有两个存储桶，(b) 其他连续数字特征将被划分为5个存储桶。

```
binary_split_cols = ['CHAS', 'RAD']
col_names = ['CRIM', 'ZN', 'INDUS', 'CHAS', 'NOX', 'RM', 'AGE',
'DIS', 'RAD', 'TAX', 'PTRATIO', 'B', 'LSTAT']
X_dtrain = {col: x_train[:, ix] for ix, col in
enumerate(col_names)}
X_dtest = {col: x_test[:, ix] for ix, col in enumerate(col_names)}

# Create feature columns!
feature_cols = []
for ix, column in enumerate(x_train.T):
    col_name = col_names[ix]
```

```
    # Create binary split feature
    if col_name in binary_split_cols:
        # To create 2 buckets, need 1 boundary - the mean
        bucket_boundaries = [column.mean()]
        numeric_feature =
tf.feature_column.numeric_column(col_name)
        final_feature =
tf.feature_column.bucketized_column(source_column=numeric_feature,
boundaries=bucket_boundaries)
    # Create bucketed feature
    else:
        # To create 5 buckets, need 4 boundaries
        bucket_boundaries = list(np.linspace(column.min() * 1.1,
column.max() * 0.9, 4))
        numeric_feature =
tf.feature_column.numeric_column(col_name)
        final_feature =
tf.feature_column.bucketized_column(source_column=numeric_feature,
boundaries=bucket_boundaries)

    # Add feature to feature_col list
    feature_cols.append(final_feature)
```

> 最好将输入函数的 shuffle 选项设置为 True 进行训练，设置为 False 进行测试。每个 epoch 都对 X 和 Y 训练集进行 shuffle，但测试期间不 shuffle。

6. 使用 TensorFlow 估计器的输入库中的 numpy 输入函数声明我们的数据输入函数。指定创建的训练观测词典和一组 y 目标。

```
input_fun = tf.estimator.inputs.numpy_input_fn(X_dtrain, y=y_train,
batch_size=batch_size,         num_epochs=10, shuffle=True)
```

7. 声明模型并开始训练：

```
model = regression_classifier(feature_columns=feature_cols,
                              n_trees=n_trees,
                              max_depth=max_depth,
                              learning_rate=0.25,
                              n_batches_per_layer=batch_size)
model.train(input_fn=input_fun, steps=train_steps)
```

8. 在训练期间，我们应该看到类似如下输出：

```
INFO:tensorflow:Using default config.
WARNING:tensorflow:Using temporary folder as model directory:
/tmp/tmpqxyd62cu
INFO:tensorflow:Using config: {'_model_dir': '/tmp/tmpqxyd62cu',
'_tf_random_seed': None, '_save_summary_steps': 100,
'_save_checkpoints_steps': None, '_save_checkpoints_secs': 600,
'_session_config': None, '_keep_checkpoint_max': 5,
'_keep_checkpoint_every_n_hours': 10000, '_log_step_count_steps':
100, '_train_distribute': None, '_device_fn': None, '_service':
None, '_cluster_spec':
<tensorflow.python.training.server_lib.ClusterSpec object at
0x7f43129d77b8>, '_task_type': 'worker', '_task_id': 0,
```

```
'_global_id_in_cluster': 0, '_master': '', '_evaluation_master':
'', '_is_chief': True, '_num_ps_replicas': 0,
'_num_worker_replicas': 1}
INFO:tensorflow:Calling model_fn.
INFO:tensorflow:Done calling model_fn.
INFO:tensorflow:Create CheckpointSaverHook.
INFO:tensorflow:Graph was finalized.
INFO:tensorflow:Running local_init_op.
INFO:tensorflow:Done running local_init_op.
INFO:tensorflow:Saving checkpoints for 0 into
/tmp/tmpqxyd62cu/model.ckpt.
INFO:tensorflow:loss = 691.09814, step = 1
INFO:tensorflow:global_step/sec: 587.923
INFO:tensorflow:loss = 178.62021, step = 101 (0.171 sec)
INFO:tensorflow:Saving checkpoints for 127 into
/tmp/tmpqxyd62cu/model.ckpt.
INFO:tensorflow:Loss for final step: 37.436565.
Out[190]:
<tensorflow.python.estimator.canned.boosted_trees.BoostedTreesRegre
ssor at 0x7f43129d7470>
```

9. 为了评估我们的模型，为测试数据创建另一个输入函数，并获得每个测试数据点的预测。以下代码能获取预测并打印平均绝对误差（MAE）：

```
p_input_fun = tf.estimator.inputs.numpy_input_fn(X_dtest, y=y_test,
batch_size=batch_size, num_epochs=1, shuffle=False)
# Get predictions
predictions = list(model.predict(input_fn=p_input_fun))
final_preds = [pred['predictions'][0] for pred in predictions]

# Get accuracy (mean absolute error, MAE)
mae = np.mean([np.abs((actual - predicted) / predicted) for actual,
predicted in zip(y_test, final_preds)])
print('Mean Abs Err on test set: {}'.format(acc))
```

10. MAE 大约为 0.71。请注意，由于 shuffle 具有随机性，你可能会得到略微不同的结果。为了提高准确度，可以考虑增加 epoch 或引入更低的学习率甚至是某种类型的衰减学习率（指数或线性）：

```
Mean Abs Err on test set: 0.7111111111111111
```

11.6.3 工作原理

本节我们介绍了如何使用 TensorFlow 的估计器和数据输入函数。这些功能非常强大，不仅使我们的 TensorFlow 代码更短和更易读，而且还提高了算法的效率并减少了创建和测试算法所需的开发时间。

11.6.4 参考

有关决策树、随机森林、梯度提升森林以及可微性、平滑性和连续性背后的数学原理，请阅读以下参考文献：

1. Tutorial on decision trees. From a machine learning crash course by Berkeley. https://ml.berkeley.edu/blog/2017/12/26/tutorial-5/
2. Random forest python tutorial. By Chris Albon. https://chrisalbon.com/machine_learning/trees_and_forests/random_forest_classifier_example/
3. A nice PDF presentation on convex functions, how they are used in machine learning, and the differences between smoothness, differentiability, and continuity. By Francis Bach. Also has ~6 pages of useful references at the end, which the reader may find helpful. https://www.di.ens.fr/~fbach/gradsto_allerton.pdf
4. Article on Soft Decision Trees: *Distilling a Neural Network into a Soft Decision Tree.* Frosst and Hinton. 2017. https://cex.inf.unibz.it/resources/Frosst+Hinton-CExAIIA_2017.pdf
5. TensorFlow implementation of a Neural Tree. By Benoit Deschamps. https://github.com/benoitdescamps/Neural-Tree

11.7 将 Keras 作为 TensorFlow API 使用

TensorFlow 为程序员提供了灵活性和强大功能，但是缺点是模型原型设计和迭代各种测试对程序员来说可能很麻烦。Keras 是深度学习库的封装器，可以更轻松地处理模型的各个方面，并使编程更容易。本节我们选择基于 TensorFlow 使用 Keras。事实上，使用带有 TensorFlow 后端的 Keras 非常受欢迎，即 TensorFlow 中有一个 Keras 库。本节我们将使用该 TensorFlow 库在 MNIST 数据集上搭建全连接神经网络和简单的 CNN 图像网络。

11.7.1 开始

本节我们将使用 TensorFlow 内部的 Keras 函数。Keras（https://keras.io/）已经是一个可以安装的独立 Python 库。如果选择使用纯 Keras，则必须为 Keras 选择后端（如 TensorFlow）。

本节我们将在 MNIST 图像识别数据集上实现两个独立的模型。第一个是前馈全连接神经网络，而第二个是复制 8.2 节的 CNN 网络。

11.7.2 动手做

1. 导入必要的编程库：

```
import tensorflow as tf
from sklearn.preprocessing import MultiLabelBinarizer
from keras.utils import to_categorical
from tensorflow import keras
from tensorflow.python.framework import ops
ops.reset_default_graph()

# Load MNIST data
from tensorflow.examples.tutorials.mnist import input_data
```

2. 在TensorFlow中加载带有MNIST数据导入函数的库。虽然原始MNIST图像像素是28×28,但导入的数据是扁平版本,其中每个观测量是0～1的784个灰度点,y标签导入了0～9的整数。

```
mnist = input_data.read_data_sets("MNIST_data/")
x_train = mnist.train.images
x_test = mnist.test.images
y_train = mnist.train.labels
y_test = mnist.test.labels
y_train = [[i] for i in y_train]
y_test = [[i] for i in y_test]
```

3. 将使用scikit-learn的MultiLabelBinarizer()函数将目标整数转换为独热编码向量,如下所示:

```
one_hot = MultiLabelBinarizer()
y_train = one_hot.fit_transform(y_train)
y_test = one_hot.transform(y_test)
```

4. 创建一个三层全连接神经网络,隐层节点数为32、16和10。输出节点数为10(每个数字对应一个节点)。网络创建代码如下:

```
# We start with a 'sequential' model type (connecting layers
together)
model = keras.Sequential()

# Adds a densely-connected layer with 32 units to the model,
followed by an ReLU activation.
model.add(keras.layers.Dense(32, activation='relu'))

# Adds a densely-connected layer with 16 units to the model,
followed by an ReLU activation.
model.add(keras.layers.Dense(16, activation='relu'))

# Add a softmax layer with 10 output units:
model.add(keras.layers.Dense(10, activation='softmax'))
```

5. 要训练模型,就要调用compile()方法,并设置参数,所需参数是优化器函数和损失函数类型。但我们也想记录模型的准确度,因此metrics列表包括accuracy参数。

```
model.compile(optimizer=tf.train.AdamOptimizer(0.001),
              loss='categorical_crossentropy',
              metrics=['accuracy'])
```

6. 输出结果类似如下:

```
Epoch 1/5
   64/55000 [..............................] - ETA: 1:44 - loss:
2.3504 - acc: 0.0625
 3776/55000 [=>............................] - ETA: 2s - loss:
1.7904 - acc: 0.3676
...
47104/55000 [=========================>....] - ETA: 0s - loss:
0.1337 - acc: 0.9615
50880/55000 [==========================>...] - ETA: 0s - loss:
```

```
0.1336 - acc: 0.9617
55000/55000 [==============================] - 1s 13us/step - loss:
0.1335 - acc: 0.9615
Out[]: <tensorflow.python.keras.callbacks.History at
0x7f5768a40da0>
```

 要为均方误差损失配置回归模型,我们将编译模型如下:`model.compile(optimizer= tf.train.AdamOptimizer(0.01), loss='mse', metrics=['mae'])`

7. 使用两个卷积层实现 CNN 模型,并且最大池化层后跟着全连接层。首先,我们必须将平面图像重塑为 2D 图像,并将 y 目标转换为 numpy 数组,如下所示:

```
x_train = x_tra0in.reshape(x_train.shape[0], 28, 28, 1)
x_test = x_test.reshape(x_test.shape[0], 28, 28, 1)
input_shape = (28, 28, 1)
num_classes = 10

# Categorize y targets
y_test = to_categorical(mnist.test.labels)
y_train = to_categorical(mnist.train.labels)
```

8. 我们将像以前一样以类似的顺序层方法创建 CNN。这次我们将使用 Keras 的 Conv2D()、MaxPooling2D() 和 Dense() 函数来创建我们的 CNN 模型,如下所示:

```
cnn_model = keras.Sequential()
# First convolution layer
cnn_model.add(keras.layers.Conv2D(25,
                                  kernel_size=(4, 4),
                                  strides=(1, 1),
                                  activation='relu',
                                  input_shape=input_shape))
# Max pooling
cnn_model.add(keras.layers.MaxPooling2D(pool_size=(2, 2),
                                        strides=(2, 2)))
# Second convolution layer
cnn_model.add(keras.layers.Conv2D(50,
                                  kernel_size=(5, 5),
                                  strides=(1, 1),
                                  activation='relu'))
# Max pooling
cnn_model.add(keras.layers.MaxPooling2D(pool_size=(2, 2),
                                        strides=(2, 2)))
# Flatten for dense (fully connected) layer
cnn_model.add(keras.layers.Flatten())
# Add dense (fully connected) layer
cnn_model.add(keras.layers.Dense(num_classes,
activation='softmax'))
```

9. 编译模型,设置优化器和损失函数。

```
cnn_model.compile(optimizer=tf.train.AdamOptimizer(0.001),
                  loss='categorical_crossentropy',
                  metrics=['accuracy'])
```

10. Keras 还允许我们将函数插入到名为 Callbacks 的训练代码中。callbacks 函数在代码中的某些时间执行，可用于执行各种函数。有许多预先定义的 callback，可以保存模型、在指定条件下停止训练、记录值等。有关不同选项的更多信息，请参考 https://keras.io/callbacks/。为了说明如何制作自定义的 callback 并展示它们如何工作，我们将创建一个名为 RecordAccuracy() 的 callback 类，它是一个 Keras Callback 类，并在每个 epoch 的末尾存储准确度，如下所示：

```
class RecordAccuracy(keras.callbacks.Callback):
    def on_train_begin(self, logs={}):
        self.acc = []

    def on_epoch_end(self, self, batch, logs={}):
        self.acc.append(logs.get('acc'))

accuracy = RecordAccuracy()
```

11. 使用 fit() 方法训练 CNN 模型。此处提供 validation_data 和 callbacks，如下所示：

```
cnn_model.fit(x_train,
              y_train,
              batch_size=64,
              epochs=3,
              validation_data=(x_test, y_test),
              callbacks=[accuracy])

print(accuracy.acc)
```

12. 训练结果类似如下：

```
Train on 55000 samples, validate on 64 samples
Epoch 1/3
   64/55000 [..............................] - ETA: 2:59 - loss: 2.2805 - acc: 0.0625
  192/55000 [>.............................] - ETA: 1:14 - loss: 2.2729 - acc: 0.1302\
...
54848/55000 [============================>.] - ETA: 0s - loss: 0.0603 - acc: 0.9816
54976/55000 [============================>.] - ETA: 0s - loss: 0.0603 - acc: 0.9816
55000/55000 [==============================] - 26s 469us/step - loss: 0.0604 - acc: 0.9816 - val_loss: 0.0139 - val_acc: 1.0000
Out[]: <tensorflow.python.keras.callbacks.History at 0x7f69494c7780>

[0.9414363636450334, 0.9815818181731484, 0.9998980778226293]
```

11.7.3 工作原理

本节我们展示了如何使用 Keras 轻松地创建和训练模型。它能自动处理变量类型、维度和数据输入的许多复杂细节。虽然 Keras 让人放心，但也应该意识到，如果掩盖太多模型细节，我们可能会无意地实现错误的模型。

11.7.4 参考

有关 Keras 的更多信息，请参考以下资源：

- Official Keras documentation: `https://keras.io/`
- TensorFlow Keras Tutorial: `https://www.tensorflow.org/guide/keras`
- "An Introduction to Keras", a guest lecture by Francois Chollet at Standford (slides in a PDF): `https://web.stanford.edu/class/cs20si/lectures/march9guestlecture.pdf`

推荐阅读